AF702529

Copyright © 2023 by Ian Allan

First edition **eBook ASIN:** B0CF2PY67W **Audiobook ASIN:** B0CKLZ51J2 **Paperback ISBN-13:** 979-8853989245 **Large Print Hardback ASIN:** B0CGL1BBPR

Copyright © 2025 by Ian Allan

Second edition **eBook ASIN:** B0DVSXFB18 **Paperback ISBN-13:** 978-0-6454465-7-9 **Large Print Paperback ISBN-13: 978-0-6454465-8-6**

Subjects: kindle paperwhite, e-reader, pdf reader

All rights reserved. Except as permitted under the Australian Copyright Act 1968 (for example, a fair dealing for the purposes of study, research, criticism or review), no part of this book may be reproduced, stored in a retrieval system, communicated or transmitted in any form or by any means without prior written permission. Enquiries via ian@ianallanauthor.com.

Disclaimer

The material in this publication is general in nature only. It does not represent professional advice. Everybody's circumstances are different. If you need expert assistance, see the service of an appropriate professional. To the maximum extent permitted by law, the author and publisher disclaim all responsibility and liability to any person, arising directly or indirectly from them taking or not taking action based on the information in this publication.

Ian Allan is not affiliated with and does not endorse any of the corporate entities mentioned in or involved in the distribution of this work, or any third party entities whose trademarks and logos may appear on this work.

The menus and techniques I've written about were correct at the time of my last edit (December 2024). However, Amazon often updates its menus, apps and website. So, some items might have changed, and the Amazon website might be different between where you live and where I live.

Contents

About Ian Allan

The Kindle Paperwhite User Guide 2025 is my 11th technology book if you include the three 4.5 star courses I teach to my 10,000 or so students on the Udemy platform.

I've been teaching technology to beginner audiences since the 1990s, and I've come to realize that too many teachers suffer from the curse-of-knowledge. That's where they assume that if something is simple for them, then it must also be simple for everybody else. That's not me. I'm good at breaking down the overwhelming into smaller manageable parts. I hope you come to agree as you read through my Kindle Paperwhite User Guide 2025.

I bought my first Kindle in 2010, my Paperwhite in 2020, my Scribe in 2023 and my Oasis in 2024. Throughout 2024, the first edition of this book was a *Best Seller* in multiple Amazon categories.

Writing the Second Edition has added even more depth to my understanding of the Kindle and the Amazon ecosystem. While my Kindle Scribe has been as revolutionary tome as a writer, my Paperwhite has been revolutionary to me as a reader. I can't wait to share what I've learnt with you.

Let's get on with it. I'll see you inside.

Ian

Chapter 1

Chapter summaries

Here's what's in each chapter. . .

Chapter 2 - Introduction: What I love about my Paperwhite, and using the Amazon ecosystem to overcome the Paperwhite's limitations.

Chapter 3 - Quick start: How to quickly setup your Paperwhite, power it up and read a book.

Chapter 4 - Setup: The Paperwhite's setup procedure step-by-step.

Chapter 5 - Tour: The different screens and toolbars, and step-by-step how to buy and read your first book.

Chapter 6 - Learning aid: Dictionary, Wikipedia, and translation functionality, as well as X-Ray, text highlighting, and typed and handwritten sticky-notes.

Chapter 7 - Screen appearance: Fonts, text size, screen orientation, screen brightness, and more.

Chapter 8 - Acquiring and managing content: How to bring both Amazon and non-Amazon content onto your Paperwhite. How to use the Kindle store, and how to send books to your Paperwhite using e-mail, the cloud and USB cable.

Chapter 9 - The Settings menu: Dictionaries, languages, accessibility, parental controls, and more.

Chapter 10 - Kindle management from a web browser: Manage your Paperwhite and read Kindle books from a browser. Review your notes and highlights, and see color versions of all your books.

Chapter 11 - Subscriptions: The different Amazon book subscriptions, and the categories of books in the Amazon store. Understanding these can help you squeeze the most out of your Paperwhite experience.

Chapter 12 - Troubleshooting: Fifteen troubleshooting problems relating to setup, content, charging and hardware.

Chapter 13 - FAQs: Twenty seven Frequently Asked Questions.

Chapter 14 - Concluding remarks

OK, let's get cracking.

Chapter 2

Introduction

I love my Kindle Paperwhite.

I love the October 2024 release. It's the fastest, most responsive Kindle yet.

I love that it's waterproof - I can read by the water and in the bath.

I love that its touch screen - the tiny buttons on my old kindle were driving me crazy.

I love that it's backlit – no more book light.

I love that it's e-ink – it's so easy on my eyes, and I can easily adjust the screen brightness and font size.

I love that I can highlight text and take notes while I'm reading – that's great for nonfiction.

I love that I can search for any word or term – in a book, in my library, or even in the kindle store.

I love that I can move seamlessly between Kindle apps and Kindle devices – Amazon's Whispersync cloud sees to that.

I love that my books are safe – if I lose my paperwhite, my books, notes and highlights are all stored in the cloud.

I love the massive bookstore that sits behind my paperwhite – most books are reasonably priced and some are even free. I read more widely and more often than I ever did.

I love that my paperwhite is lightweight and slim - it easily slips into the front pocket of my bag.

I love its familiar interface. Much like windows and apple computers, all my Kindle's have near-identical navigation.

I love that my paperwhite is a reading tool that knows its place - it does what it's meant to do, and it does it well.

Yes, there's a lot to love about my paperwhite. Did I mention that I love my paperwhite?

I'm going to talk about all those things, but for now, enough of the prose!

2.1 How to make the Amazon ecosystem work for you

Much like some people use a car for Sunday drives and other people are always searching for a car's turbo-boost button, your Paperwhite can be used for pleasure-reads and occasional note taking. Or, by harnessing Amazon's ecosystem, it can become a powerful and flexible reading and writing tool. In a nutshell, those two extremes are what this book is about.

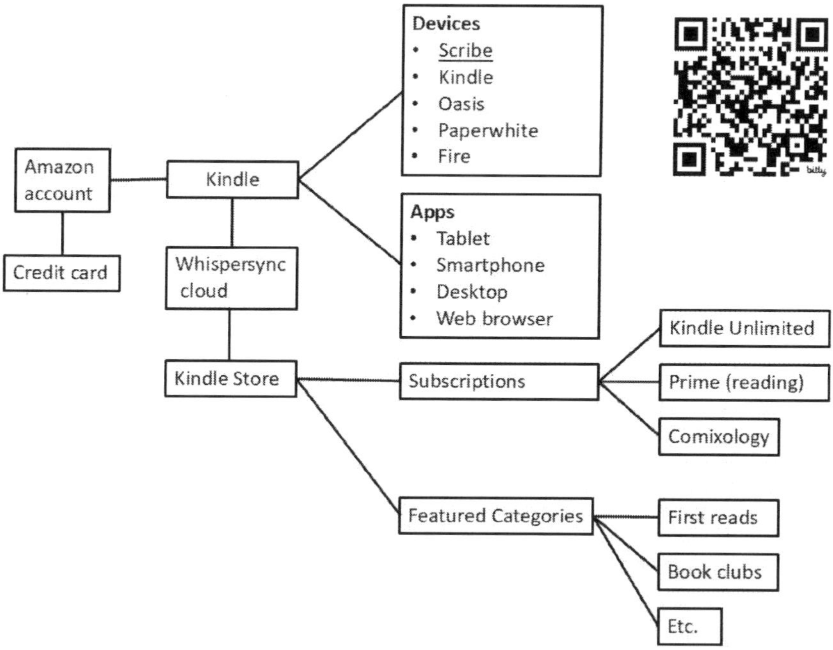

Figure 2.1: Diagram of Amazon's ecosystem.

Amazon’s ecosystem relies on you being logged into your account so it can curate content for you, and synchronize all your books across all your Kindle devices. Download the figures from bit.ly/akpug or the QR code.

Let’s start with a 30,000-foot overview of Amazon’s eBook ecosystem. By ecosystem, I mean, enabled by the cloud, everything that’s linked to your Amazon account - subscriptions, content and device management from your web browser, and synchronized purchasing, reading, highlighting and note taking functionality across your Paperwhite and other Kindle apps and devices. If you're a bit overwhelmed by that last sentence, that's because there's a lot to this deceptively simple ecosystem. Let’s look at its five parts . . .

2.1.1 Ecosystem part #1 - Subscriptions

If you read two or more books a month, I strongly encourage you to consider becoming a Prime member or a Kindle Unlimited subscriber. Prime members have thousands of books to choose from. Kindle Unlimited subscribers have millions of books to choose from. Some books even come bundled with narration. Prime Reading tends to be a high quality subset of Kindle Unlimited books. More on subscriptions in chapter 11.

2.1.2 Ecosystem part #2 - Book recommendations

Amazon uses your reading and searching history to curate lists of book recommendations. By your 'reading and searching history' I mean that Amazon uses every logged in device (Kindle devices, Kindle apps, web browser), to track and synchronize your reading and searches.

I like to read widely so I find it frustrating that Amazon usually recommends more of the same genre's and authors that I (and my partner who uses the same account) have been reading! However, I've learnt that the Kindle app, Amazon store in a web browser, and Kindle devices (Paperwhite, paperwhite, etc) have different ways of showing the recommendations. If you like to read widely, you can maximize Amazon's range of recommendations by alternating between the kindle stores on your Kindle device, the apps (eg. on your smartphone or desktop), and the Amazon store in a web browser. If you only read one genre then Amazon will continue to mostly recommend that genre to you.

The best way to get un-curated book recommendations is to log out of Amazon and search the store from a browser in private mode.

2.1.3 Ecosystem part #3 - Content and device management from a browser

You can manage your books (what's on your Paperwhite, every Kindle app, and every kindle device you own) from a web browser. You can also view your notes and highlights, and purchasing and borrowing histories (chapter 10).

2.1.4 Ecosystem part #4 - Synchronized reading

Thanks to Whispersync, your purchases, reading, note taking and highlighting are synchronized across all your apps and devices in near-real time. There's a Kindle app for most smart phones, tablets, and desktops. You can also read Kindle books in a web browser (chapter 10).

2.1.5 Ecosystem part #5 - Overcoming your Paperwhite's limitations

The Paperwhite is an amazing e-reader and note-taker, but beyond that, it can be clunky. The good news is that, enabled by Amazon's ecosystem, you can use the Kindle app, a browser or kindle device that's best suited to your task. Let me elaborate. . .

- **Audio:** Some Kindle books come bundled with audio. The Kindle apps are much better suited to audiobook listening than your Paperwhite (chapter 9).

- **Note taking while you're reading a book:** I do basic note taking and highlighting when I'm reading (chapter 6). I often view and edit them in a web browser on my laptop. I can also view them on any Kindle app or device (chapter 10).

- **Account management:** You can only do basic account management on your Paperwhite. A web browser is much better suited to that (chapter 10).

2.2 Getting the most out of this book

It's best to have your Paperwhite with you while you read this book. Some of the topics may seem a little abstract otherwise. Please download the searchable/printable PDF of all the figures from bit.ly/akpug (free and no signup required), and have them handy. Here's some ways you might do that...

- Use sendtokindle (section 8.2.1) to send my searchable PDF to a second device and display the figures there, or
- Print the figures out, or
- Open this book on a second kindle app or device and display the figures there, or
- Bookmark the figures on your Paperwhite (label 33 in Figure 5.4) and access them as required.

Also, be sure to use the Table of Contents, or use the search bar (label 34 in Figure 5.4) to find topics that interest you.

Whenever I can, I include the path to a menu in italics below each section heading. Along with describing the menus, I explain how to use them and why they matter. For tasks your Paperwhite doesn't handle well, I suggest alternatives.

2.3 Chapter guide – Choose your own adventure

Here's some ideas for how different people might make the most use of this book . . .

- **If you just want to get started, look at. . .**
  - The 5-minute QuickStart in chapter 3.
  - Detailed setup in chapter 4.
  - Buying and reading your first Kindle book in chapter 5.
  - Troubleshooting screen taps in chapter 12.
- **Amazon's ecosystem. Everybody should understand this!**
  - Note taking in chapter 6.
  - Buying books on your smartphone and transferring books from your computer in chapter 8.
  - Managing your Paperwhite from a web browser in chapter 10.
  - Amazon subscriptions explained in chapter 11.
- **If you're a language learner, look at. . .**
  - Inbuilt dictionary, X-Ray and vocabulary builder in chapter 6.
  - Continuous word definitions and Word Wise in chapter 7.

   - Setting up languages, translations and dictionaries in chapter 9.
   - Books with narration in chapter 11.
- **If you're a nonfiction reader, look at. . .**
   - Note taking and highlighting on your Paperwhite in chapter 6.
   - Note taking and highlighting in the kindle cloud reader in chapter 10.
   - Organizing your books into collections in chapter 8.
- **If you need the accessibility features, look at. . .**
   - Font size and dyslexic font in chapter 7.
   - Voice view screen and menu reader in chapter 9.

Chapter 3

Five minute setup and overview

So, you've taken delivery of your Paperwhite and you just want to get started. In this chapter I walk you through the fastest way to set up your Paperwhite and use it for the first time. For most people, the setup is trouble free and intuitive, and they're up and running in less than five minutes.

In Chapter 4, I go into setup in much greater detail, including troubleshooting. Hopefully you won't need to look at that chapter. For now, let's focus on getting your device charged, powered up, and connected to Wi-Fi. Ready? Let's go!

3.1 Charging

You charge your Paperwhite from the port at its base, preferably using the USB-C cable that came with it. It will charge from most USB chargers. For me, that's my iPhone charger, computer port and the USB port in my car. An amber light near the charging port means that it's charging, and a green light means that it's fully charged.

3.2 Powering up

The power button is at the base, an inch to the right of the charging port. Press it until you see a green light and your Paperwhite will power up.

3.3 Setup

If you bought your Paperwhite directly from Amazon and ticked the link to your account box when you checked out, then you will only need to connect to your Wi-Fi from the Wi-Fi list that pops up during setup. However, if you bought your Paperwhite from your local retailer, you'll be stepped through a setup process that also includes signing up to the goodreads book review site and signing into your Amazon account.

3.4 First use

When setup is complete, you will be taken to the Home screen (Figure 5.1). Notice the three items at the bottom. Home on the left, a book thumbnail in the middle, and Library on the right.

- **Home:** Tap Home on the left side of the bottom menu bar to be taken to the Home screen. The Home screen includes zones such as From your library – they're the most recent books that you've accessed. There's also a bunch of other categories that Amazon curates for you based on what it knows about your reading preferences.

    - If you tap on a book in the Home screen, the book will open if you own it. Otherwise, you will be taken to the book's sales page where you can view more information about it or buy it. Or, if you're a Prime or Kindle Unlimited subscriber, borrow it (if it's an eligible book).

- **Book thumbnail:** In the middle of the bottom menu bar is a thumbnail of the most recent book that you have been reading, or notebook that you've been working on. Tap to open it.

- **Library:** Tap Library, second on the left side of the bottom menu bar, to be taken to the Library screen. Your library has thumbnails of all the books you own, or have borrowed from Prime Reading or Kindle Unlimited. If you are new to Kindle, this area might contain a couple of free books, compliments of Amazon. When you tap on a book thumbnail, the book will download and open.

3.5 Finding your way around an open book

In an open book (Figure 5.2). . .

- Pinch two fingers together or drag them apart to change the text size.

- Tap the right side of the screen to go to the next page.

- Tap the left side of the screen to go to the previous page.

- Swipe down from the top of the screen to see the screen warmth slider, and screen brightness slider.

- Tap the top of the screen to reveal the Standard toolbar. Tap the back-arrow in the top left corner to exit the book or notepad, and be taken back to either the Home screen or Library screen.

3.6 Some icons are contextual

By contextual I mean that they link to menus that are relevant to the task you are doing at the time. For example, the [three vertical dots] icon launches different menus from your Home screen, Library screen and Notebooks screen, from within a book or notepad, and from pop-up menus during different tasks.

Context also comes into play when you tap, hold and release on a word, phrase, or passage. If you've selected a noun in an X-Ray enabled book, the X-Ray menu pops up and you can find all the passages in the book where the noun is mentioned. Selecting any other word launches the in-built dictionary, as well as options to see it's Wikipedia entry or the translation menu. Or, you could just highlight a passage and use the keyboard that appears to write notes about it.

3.7 The Paperwhite has weaknesses

For all the things I love about my Paperwhite, it does have its weaknesses. Fortunately, these tend to be strengths in the Kindle apps. For example, the keyboard is a little clunky when it comes to taking notes. However, your notes are easy to revise (and even color-code) in any one of the kindle apps.

3.8 Your amazon account in a browser

From your local amazon web page, add /myk to the URL (eg am azon.com/myk amazon.co.uk/myk)

In a browser, be sure to familiarize yourself with the Content and Devices area of your amazon account. That's where you can look at your purchasing, and Prime and Kindle Unlimited borrowing history, and set up email permissions for transferring non-amazon content to your Paperwhite.

Well, that should be enough to get you up-and-running. Like many things in life, though, the devil is in the detail. Read on and we'll get into the nuts and bolts.

Chapter 4

Detailed setup

In this chapter I walk you through all the setup scenarios I could think of. Whether you bought your Paperwhite directly from Amazon, bought it in a store, or are troubleshooting a tricky connection, you'll find the help you need here.

If you've already managed to connect your Paperwhite to Wi-Fi and your Amazon account, and you can open a book, then you should skip to the next chapter. Otherwise, read on to be shown what to expect when you power up your Paperwhite for the first time. There are three setup scenarios. . .

- **Setup scenario 1 - Your Paperwhite was setup by amazon at checkout:** If you bought your Paperwhite directly from Amazon and clicked on the Link to my account box at checkout, you will only need to charge it, power it up, and connect it to Wi-Fi (chapter 3).

- **Setup scenario 2 - First use auto setup:** If you bought your Paperwhite from somewhere other than Amazon, when you power up, your Paperwhite will guide you through a 7 step setup process.

- **Setup scenario 3 - Manual setup:** You go down this path if the auto-setup did not work for you.

The days of needing a technical person to set up a device for you are mostly gone. Setting up a Paperwhite is a trouble-free exercise for most people. But there can be problems. For some reason, that's usually me! If at the end of this chapter, you're still not setup, chapter 12 includes a bunch of troubleshooting tips.

4.1 Before you begin

There are three things that you should do before you set up your Paperwhite.

- **Charging:** Make sure your Paperwhite is fully charged so that it does not unexpectantly power down during setup and potentially corrupt the process.

- **Wi-Fi:** Make sure you are in range of Wi-Fi and that you have your Wi-Fi password handy.

- **Amazon account:** Have your Amazon account details handy. If you don't already have an account, in a web browser, go to your country's amazon store and create one.

4.2 Charging your Paperwhite

Quickstart: Plug the cable that came bundled with your Paperwhite into a power source such as a phone charger or computer. An amber light near the charging port shows that it's charging. A green light shows that it's charged.

Your Paperwhite comes with a USB cable for charging from either your computer or wall adapter. To ensure trouble-free setup, it's important to fully charge it first.

A full charge takes 7 hours when connected to a computer, 2.5 hours with a 9W USB charger. The USB-C cable that came with your Paperwhite is likely to work better than no-name cables.

An amber light near the charging point means that it's charging, and a green light means that it's fully charged.

You can use your Paperwhite while its charging. A lightening bolt symbol appears in the top right corner of the screen. If you're charging from your computer, you should unmount / eject the kindle drive to avoid corrupting its memory. Tap the Disconnect button that appears on your Paperwhite's screen.

See chapter 12 to troubleshoot charging problems.

4.3 The power button

Quickstart: To power on and power off, press the power button until you see a green light. The button is around an inch to the right of the charging port. The power-light is around an inch to the left of the charging port.

The power button at the base of your Paperwhite is used to start it, place it into sleep mode, turn it off, and restart it. It can be hard to see - it's about an inch to the right of the charging port.

4.3.1 Powering up

Press the power button and release it when you see your Paperwhite power up. A green light to the left of the charging port will flick on briefly.

4.3.2 Sleep mode

Place your Paperwhite into sleep mode with a quick press-and-release on the power button. Press-and-release again to wake it up.

Sleep mode begins automatically after 10 minutes of inactivity. The backlight will dim, your book will close, and the screen saver will display. Whispersync communication with the Amazon Cloud remains active so your Paperwhite will still consume some power.

You can also toggle sleep mode from the settings menu (see Power Saver in chapter 9).

4.3.3 Shutting down

Depress the power button for 10 seconds. The green light at the base will flash. Then the amber light. From the box that appears on the screen, tap Screen Off. Shutting down your Paperwhite turns off all power consuming activity, including the backlight and Whispersync.

4.3.4 Restarting

[three vertical dots in top right corner] > *settings* > *device options* > *Restart*

OR

Hold the power button down until the Power dialog appears and tap Restart

Restarting your Paperwhite is much like a reboot on your desktop computer. You may need to restart if your Paperwhite freezes, or its response times become slow. It's good practice to restart any device occasionally anyhow. There's three ways to restart your Paperwhite. . .

1. Depress the power button and continue holding it in after it has powered off (allow 40 seconds).

2. Depress the power button for 10 seconds. The green light at the base will flash. Then the amber light. From the box

that appears on the screen, tap Restart.

3. From the [3 vertical dot] menu on the top right of the screen, tap restart in the Settings > Device options path

4.4 Auto Setup

On its first use, your Paperwhite will walk you through a setup process. To avoid setup issues, make sure its fully charged before you begin. Press the start button at its base. If you bought your Paperwhite from your Amazon account and ticked the Link to my account box at checkout, you will only need to connect to Wi-Fi. Otherwise, expect the following prompts. . .

1. Select a language.

2. Select a region.

3. Connect to Wi-Fi (section 4.5).

4. Login to your amazon account using your username and password (section 4.6).

    ◦ If you don't already have an account, in a web browser, go to your local amazon store and create one.

5. Send a kindle app link to your smartphone.

    ◦ Type in your phone number and tap Send Link. Amazon will send you a text with a custom link to install the kindle app on your smartphone. When you use this link, the app will be automagically linked to your amazon account and synced to your Paperwhite.

6. Choose to create / sign in to a goodreads account using

your Amazon details.

7. Choose an audible free trial if you'd like to.

8. Setup complete.

If you accidentally chose a foreign language and you can no longer read the on-screen setup instructions, don't panic (well, maybe a little). You will need to reset your Paperwhite. I show you how to do that in chapter 12.

4.5 Connecting to Wi-Fi

Home / Library > [3 vertical dots] > Settings > Wireless > Wi-Fi Networks

Quickstart: When prompted, select a Wi-Fi network and enter a password.

If you're setting up a brand new out-of-the-box Paperwhite, you'll be prompted to select a Wi-Fi network to connect to and enter a password. If it connects successfully, go to the Sign into Amazon section below. Otherwise, read on.

Assuming you're in range of an active Wi-Fi network, tap the [3 vertical dot] icon (label (2) in Figure 5.1) and select Settings from the menu that appears, tap Wireless, and then tap Wi-Fi Networks. A list of available Wi-Fi networks will display, including a mobile phone hotspot if one is active. Select the network you want to connect to and enter a password when prompted. A confirmation message will display on the screen if the connection is successful.

You only need to connect to each wireless network once. Your Paperwhite will automatically reconnect the next time you're in range.

4.6 Sign into Amazon

[three vertical dots] > Settings > Your Account > Register Your Kindle

Quickstart: If you bought your Paperwhite directly from Amazon and clicked on the "Link to my account" box at checkout, your Paperwhite is already signed into your account.

As a part of auto-setup, a Register your kindle box appears in the welcome screen. If it does not appear, tap the [three vertical dots] icon (label (11) in Figure 5.1) and select Settings. Navigate to the Your Account menu, and then tap Register Your Kindle (chapter 9).

You could use your Paperwhite without registering it on your Amazon account, but you would have to transfer all your books using a USB cable (chapter 8). Expect that experience to be cumbersome and frustrating. Also, you would only be able to read non-amazon books. The Digital Rights Management (DRM) security that is attached to most Kindle books prevents them from being read on an unregistered Kindle.

Could you spare a moment to review my book?

Thank you so much for reading my book! If you've found it helpful, I'd be incredibly grateful if you could leave a short review. Reviews are a huge help to indie authors. They help other readers discover our books and keep us writing. Just a few words about what you liked would mean the world. Thanks again for your support!

Review on Amazon US		Review on Amazon UK	
https://amzn.to/4hFb80w Click the Flag or Link	Scan the QR code	https://amzn.to/3EA7u9K Click the Flag or Link	Scan the QR code
Review on Amazon Canada		**Review on Amazon Australia**	
https://amzn.to/3Egus3q Click the Flag or Link	Scan the QR code	https://amzn.to/4hmKEkv Click the Flag or Link	Scan the QR code

Don't forget the resources from the QR codes in the figures, and at ianallanauthor.com/paperwhite

Chapter 5

Paperwhite tour

So, your Paperwhite is setup and ready to use. In this chapter I'll show you how to find your way around the Home screen and the Library screen. Also, the menus that become available when you open a book, and how to buy your first book and read it.

5.1 Quickstart

- ***In the Home screen: Tap a book thumbnail and get taken to the book's sales page in the Kindle store.***
- ***In the Library screen: Tap a book thumbnail and it will open.***

When you're reading a book . . .

- ***Tap the right side for the next page.***
- ***Tap the left side for the previous page.***
- ***Swipe down from the top of the screen to see the Quick Access menu. That's where you'll find the screen brightness and warmth sliders. Tap the up arrow at the bottom of the menu to exit the dialog.***
- ***Tap the top of the screen to see the Standard Toolbar.***

Then tap the arrow in the top left corner to return to the Home screen or Library screen.

- ***Use a pinching action to reduce and enlarge the font size.***

5.2 The Home screen and Library screen

Figure 5.1: The Home screen and Library screen (download the figures from bit.ly/akpug or the QR code).

Your Paperwhite's Home screen and Library screen are broken into three sections (Figure 5.1). Their top and their bottom are identical. The bit in the middle is different. The Home screen is mostly a kindle store that has been curated for you based on what amazon knows about your reading preferences. The Library screen displays all your books, some or all of which will have already downloaded to your Paperwhite. Other books might still be in the cloud, so you may need to tap on their thumbnails to force them to download.

In Figure 5.1, the top menu bar contains. . .

Label (1) Search bar: Type a term into the search bar and amazon searches for it. . .

- In the titles of books within your library.
- In the titles of books in the kindle store.
- In the titles of books in goodreads.
- Inside all the books that are in your library.
- Within Wikipedia and your Paperwhite's dictionary.

Tap on a result to see the details.

Label (2) [3 vertical dots]: Tap this icon to see the following tappable menu items. Your Reading Lists (samples you've downloaded), Goodreads, Web Browser, Settings, Legal, and Create a Collection.

Label (3) Kindle store: Tap the shopping trolley icon to go to the kindle store.

Label (4) Filters: **Status** (Downloaded, Unread, Read), **Type** (Books, Documents, etc), **Your programs** (Kindle Unlimited [borrow count])

Label (5) Create a Collection: When you're in the Library tab and you choose Collections from View options (label (6) in Figure 5.1), a + icon appears. Tap the + icon to create a collection (section 5.2.2 and section 8.5).

Label (6) Sort and View options (Library tab): **Sort order** (recent, title, author, date, ascending, descending), **View options** (All, Collections, Grid view, List view)

The bottom menu bar has links to. . .

Label (7) The Home screen.

Label (8) The Library screen.

Label (9) Thumbnail of the most recent book or thumbnail you accessed.

5.2.1 Home screen

The Home screen (left screen capture in Figure 5.1) has two types of thumbnails...

- **Thumbnails in the *From your library* area:** These are books, documents and notebooks that you have opened recently.
- **Thumbnails in the *Recommendations* area:** This is a curated version of the kindle store that has been personalized for you. There's groups of thumbnails that are designed to help you choose your next book. Each group contains personalized book recommendations based on your reading history, search history and preferences. Tap on a heading to expand the recommendations. Depending on the popularity of the category, there might be just a few, or up to 99 book recommendations.
    - So, if you read lots of Romance, Amazon will recommend mostly Romance books. If you read self-help, Amazon will recommend mostly self-help books, etc. I like to read widely so I often check the amazon app and amazon store for recommendations as well. You'll get even broader recommendations if you log out of the amazon store and then search it from a browser that's in private mode.
- **Privacy:** Another option would be to toggle the privacy setting in the Privacy and Security area to OFF, so that amazon does not use your search behavior for marketing

purposes (section 9.3.5).

5.2.2 Library screen

The Library screen (right screen in Figure 5.1) is where you access and manage all of the books and other content that you have purchased or downloaded onto your Paperwhite. Just above the book thumbnails, there's a Filter icon and a Sort by / View options icon.

Label (4) The Filter icon: The Filter icon (label (4) in Figure 5.1) is on the left near the top of the Library screen. From here you can choose to Filter your library content by a combination of. . .

- **Status:** Downloaded, unread, read.
- **Type:** Book, sample, document, newsstand, comic, collection.
- **Program:** Kindle Unlimited, Prime, Audible, Comixology.

Label (5) The Create a Collection icon: When you choose to view Collections in the Sort by and View options menu (label 6), the Create a Collection + icon appears (label (5) in Figure 5.1). Tap this and the view will change from book thumbnails to collection thumbnails. A collection is like a folder in MS Windows. They're a great way to organize your reading. For example, you could have a folder called fiction and sub-folders called romance and crime. Tap the three vertical dots at the bottom right of a book's thumbnail to add it to or remove it from a collection. More on Collections in section 8.5.

Label (6) The Sort by and View options icon: The Sort by and View options icon (label (6) in Figure 5.1) is on the right near the top of the Library screen.

- **Sort by options:** You can sort your Library content by a combination of. . .
    - Most recently read, title, author, or publication date.
    - Ascending or descending.
- **The View options:** You can choose to display your content as either a grid or a list. The title and cover thumbnail of each item is displayed.
- **View as a Grid:** When displayed as a grid, the top-right corner of each thumbnail has information about your relationship to the book – "Read", "Sample", and Percentage progress eg. "50% complete". The bottom left corner shows the relationship between the book and your Paperwhite - a tick means that it's downloaded. If the book is one of a series, a number on the thumbnail indicates how many of the series you have on your Paperwhite. Or, if it's a collection you're looking at, a number on the thumbnail indicates how many books there are in the collection.
- **View as a List:** When displayed as a list, the area under the author name has information about your relationship to the book – "Read", "Sample", Percentage progress eg. "50% complete". The bottom right corner of the book's row shows the relationship between the book and your Paperwhite - a tick means that it's downloaded.
- **View Collections:** You can also choose to display your collections. I talk about collections in section 8.5.

In any of the views, you tap on a book thumbnail or a document thumbnail to open it. Sometimes your tap will trigger it to download from the cloud first. If you tap-and-hold on a thumbnail, or tap the three vertical dots to the right of a thumbnail, you'll launch the book's options menu. There's a bunch of things you can do in there - delete it, add it to a collection, view it in the Kindle Store, and more.

5.3 Reading a book

Figure 5.2: The Paperwhite's Swipe zones and Tap zones (download the figures from bit.ly/akpug or the QR code)..

The labels are described below. Swipe down from the top of the screen to see the Quick actions toolbar (Figure 5.3). Tap the top of the screen to see the Standard toolbar (Figure 5.4). Tap (book) or Swipe (pdf) on the left side of the screen for the previous page, and the right side of the screen for the next page. Tap the bottom left of the screen to toggle through your reading progress information. Download the figures from bit.ly/akpug or the QR code.

Tapping a book's thumbnail – that's any thumbnail in your Library, or the book icon at the bottom of your Home screen and your Library screen (label (7) in Figure 5.1), opens the book and takes you to the last page you read in that book. In an open book, there's TAP zones and SWIPE zones. These are labelled in Figure 5.2. . .

Label (12): Swipe down from the top of the screen and the Quick Actions toolbar will appear (Figure 5.3). Tap at the top of the screen and the Standard toolbar will appear (Figure 5.4).

Label (13): In a book, tap on the left of the screen and return to the previous page. In a PDF file, you'll need to swipe.

Label (14): Tap the bottom-left of the screen and toggle between the following reading progress options – page number, time left in chapter, time left in book, location, and blank. You can also do this from the Reading Progress menu (section 7.4.1).

Label (15): In a book, tap on the right of the screen and advance to the next page. In a PDF file, you'll need to swipe.

5.4 Read a graphic novel

To read a graphic novel you'll need to adopt a slightly different reading technique. Graphic novels have panel view and frame view options.

- **Panel view:** The panel view is a whole comic book page from where you can tap on the right for the next page or tap on the left for the previous page (see label (13) and label (15) in Figure 5.2).

- **Frame view:** If you double-tap on a frame in a panel

view, the frame will enlarge to be full-screen. You can then progress through the comic frame-by-frame by single-tapping on the right to advance to the next frame, or on the left to return to the previous frame (see label (13) and label (15) in Figure 5.2).

5.5 Zoom and pan images

You can zoom and pan on an image:

- **To enlarge an image:** Tap-and-hold on the image. Then tap the magnifying glass that appears, and the image will enlarge.
- **Zoom in or out on an image:** Spread two fingers apart to zoom in, or pinch two fingers together to zoom out. That's a useful trick for viewing the figures in this book.
- **Pan an image:** To pan an image, swipe your finger across the screen in the direction you want to move it.

5.6 The Quick Actions toolbar

Book > swipe down from top for the Quick Actions toolbar

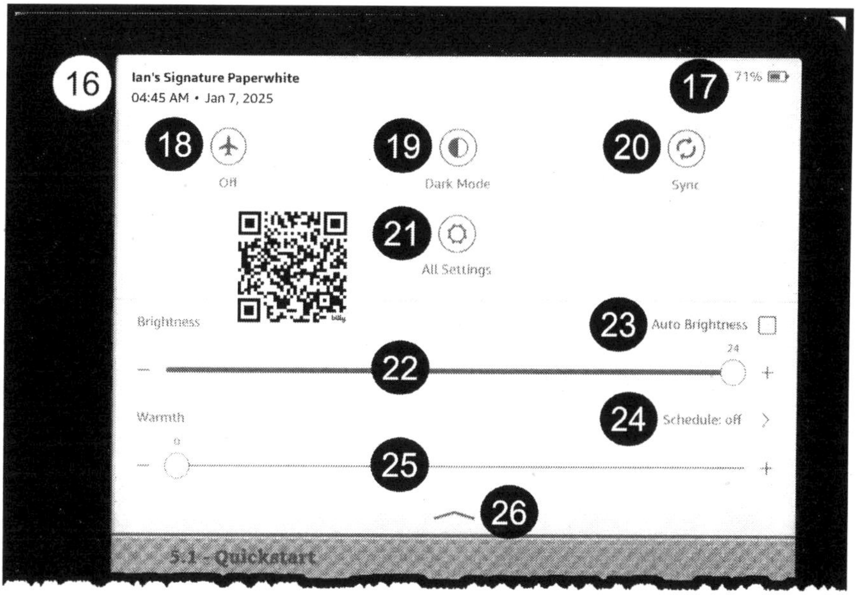

Figure 5.3: The Quick actions toolbar (download the figures from bit.ly/akpug or the QR code).

Swipe down from the top of any screen (Home, Library, or in a Book) to access the Quick Actions toolbar. The icons in Figure 5.3 are labeled so that they are easy to refer to. There's a bunch of information and tools in here. . .

Label (16) General information: Name of kindle device, date and time.

Label (17) Percentage charged: A lightening bolt appears in the middle of the battery symbol when your Paperwhite is turned on and its charging.

Label (18) Airplane mode: Tap the icon to toggle airplane mode on and off. Airplane mode can also extend the life of a battery charge because it turns off Wi-Fi.

Label (19) Dark mode: Tap the icon to toggle between black text and white background, and white text with black background.

Label (20) Synchronize: Tap the icon to manually synchronize your Paperwhite with all your kindle apps and devices. You might use this if your Scribe doesn't notice that you've bought a book on another device (see also Content problems in chapter 12).

Label (21) All settings: Tap the icon to access the Settings menu (chapter 9).

Label (22) Screen brightness slider: Tap and hold the slider, or tap the + or the - symbols for brightness values ranging from 0 to 24.

Label (23) Auto brightness: Tap to allow your Paperwhite to automatically adjust the screen brightness to different situations (eg. reading under a tree versus reading in full sunlight). I show you how to calibrate this in chapter 12.

Label (24) Schedule: Toggle the warmth levels according to a schedule. Automatically by sunrise and sunset times for your location, or manually between two times you choose.

Label (25) Warmth slider: Tap and hold the slider, or tap the + or the - symbols for warmth values ranging from 0 to 24.

Label (26) Up arrow: Tap the up arrow to minimize the Quick access menu.

5.7 The Standard toolbar

Book > tap top for Standard toolbar

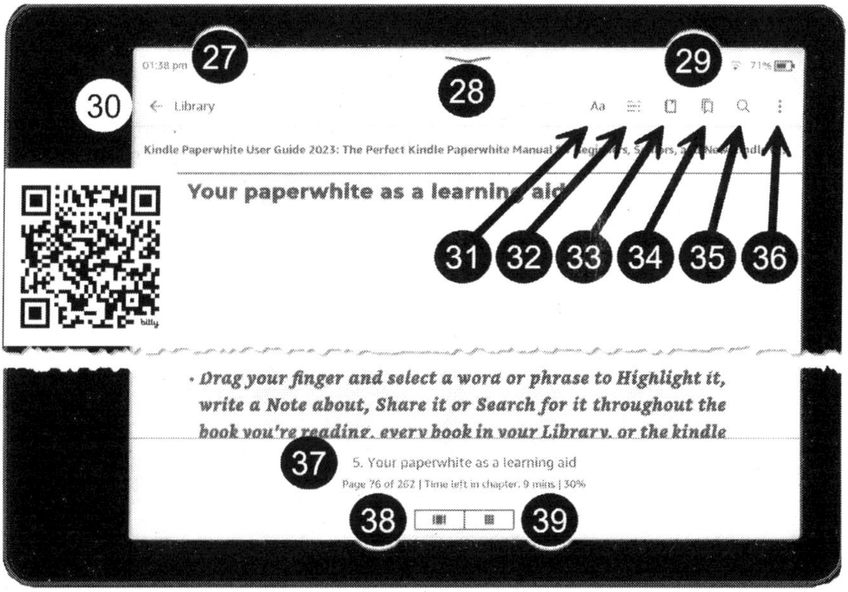

Figure 5.4: The Standard Toolbar (download the figures from bit.ly /akpug or the QR code).

While in a book, tap the top of the screen to access the Standard toolbar. There's a bunch of information in here, and there's also tools to turbo-charge your Paperwhite experience. If you are new to the Paperwhite, the most useful item in this toolbar is likely to be the Display Settings icon (label (30) in Figure 5.4 and all of chapter 7).

Label (27) Time: You set the time in the Settings menu (chapter 9). Choose between a 12 hour clock or a 24 hour clock.

Label (28) Quick Actions toolbar: Swipe down on the arrow to launch the Quick Actions toolbar (Figure 5.3).

Label (29) Wi-Fi signal strength and % charged: Wi-Fi strength shown as bars, and percentage charged.

Label (30) Return arrow: Tap this icon to return to whichever one of the Home or Library screens you were last at.

Label (31) Display Settings: Tap this icon to access the Themes, Font, Layout, and More menus (chapter 7).

Label (32) Contents and popular highlights: Tap the icon to be taken to the book's table of contents page and Notes and Highlights page (chapter 6).

Label (33) Notes and highlights: Handwritten notes, typed notes and highlights (chapter 6 and section 10.2).

Label (34) Bookmarks: Tap this icon to manage your Bookmarks (chapter 6).

Label (35) Search: Tap this icon to launch the search bar and keypad. You can search for a word in your book, and in your notes and highlights.

Label (36) [3 vertical dots]: Tap this icon for additional options - information about the book, toggle to show or hide the Writing toolbar, X-Ray and Vocabulary builder (chapter 6), the Settings menu (chapter 9), Sync, and Share a reference to your book (chapter 6).

Label (37) Reading progress: This area displays information about your reading progress, where you are in the book, and how much of the book you have read.

The Pageflip toolbar allows you to preview a different page in a book without losing your place.

Label (38) Pageview: Tap this icon to display Pageflip as a single page.

Label (39) Grid view: Tap this icon to display Pageflip as a 9 page grid. Tap the X in the top right corner of the box to return to your book.

5.8 The kindle store

Your Paperwhite is part of the Amazon ecosystem. It is linked to your amazon account, and the massive Kindle Store that contains millions of titles, including bestsellers, new releases, and classic literature.

Due to the royalty sharing arrangements with authors (books less than $10 attract double the royalty), Kindle books are often a fraction of the cost of their hardcopy cousins, and they're available near-instantly.

In this section I'll show you how to use the kindle store from your Paperwhite. You can access it from . . .

- The curated kindle store on your Home screen (Figure 5.1).
- The search bar on your Home screen and your Library screen (label (1) in Figure 5.1).
- The shopping cart button on your Home screen and your Library screen (label (3) in Figure 5.1).

There's other ways to buy and send books to your Paperwhite too. I talk about those in chapter 8.

5.8.1 Kindle Store from the Home Screen

The Home screen (Figure 5.1), is a version of the kindle store that's customized to you. It's divided into categories like From your Library, Kindle Unlimited Recommendations, Similar to your last read, etc. Tap the > symbol next to a heading and it will expand to a list of tappable book thumbnails and titles.

Tap a book thumbnail and you'll be taken to its sales page in the kindle store.

5.8.2 Kindle Store from the Shopping Cart icon

Home | Library > shopping cart

From the Home screen or Library screen, tap the shopping cart icon (label (3) in Figure 5.1) to open the Kindle Store. You'll be taken to a screen that looks similar to the Home screen, except that it has more options.

You'll see featured categories like Prime Reading and Kindle Unlimited (chapter 11). And under each of these categories, sub-categories like nonfiction, romance and graphic novels.

You can refine your search by categories, or scroll through the categories and books that Amazon recommends to you, based on your search and purchasing history. Tap the See more label next to the category name to expand it into a scrollable view.

5.8.3 Kindle Store from the Search Bar

Home | Library > Search Kindle

From the Home screen or Library screen, tap in the Search Kindle box (label (1) in Figure 5.1). Use the keyboard that appears, use your finger or Paperwhite pen to type in the name of a book, author, or topic, and Amazon will display a list of matching results from the Kindle store, as well as titles in your Library, goodreads, within books in your library, and the dictionary and Wikipedia.

5.8.4 Book sales page

Tap on a book thumbnail to open its sales page in the kindle store. There you can read its description, reviews, download sample pages, and buy it (chapter 8). Or, if you're a Prime member or Kindle Unlimited subscriber, you could borrow an eligible book (chapter 9).

5.9 How to Bookmark a page

Book > Tap top for Standard toolbar > Bookmark icon > icon

To manage bookmarks on your Paperwhite, while reading a book, tap the top of your screen to show the Standard toolbar, then tap the Bookmark icon (label (34) in Figure 5.4). Or, tap the top-right corner of the screen. An options box will appear so you can. . .

- **Add a bookmark:** Tap the icon.
- **View your bookmarks:** A list of all your bookmarks and their page numbers are displayed. Tap the bookmark you want to go to.
- **Delete a bookmark:** From the list of bookmarked pages, tap the bookmark, and then the that appears next to it.

5.10 Find a Page or Location

Book > tap top for Standard toolbar > Contents | Notes & Highlights icon > Page or Location.

To go to a page or a location while you're reading a book, tap the top of your screen to show the Standard toolbar, then tap the Contents icon (label (32) in Figure 5.4). Then tap the Page or Location item in the Contents box. In the box that launches, enter a page number or location number to be taken there.

Kindle books are reflowable text. That's text that you can adjust to suit your reading preferences (chapter 7). That means that the page numbers that are displayed on your Paperwhite may not match the book's printed version. So, if you need to reference a quote from an eBook, you should consider referencing a location number as well as a page number.

5.11 Table of Contents

Book > tap top for Standard toolbar > Contents | Notes & Highlights icon

To show a book's table of contents, while reading a book, tap the top of your screen to show the Standard toolbar, then tap the Contents icon (label (32) in Figure 5.4). Then tap on a chapter to be taken there.

Chapter 6

Your Paperwhite as a learning aid

In a book, tap, hold and release to select a word, phrase or passage, and a world of possibilities open up. You can highlight, annotate, note take, translate, look up a dictionary, and improve your vocabulary. Sorry, but I couldn't resist rhyming that!

6.1 Quickstart

Tap and hold on a word or passage to launch a menu bar, a box, or depending on context, both.

- ***Tap, hold and release on a word and a box will launch with Dictionary, Wikipedia and Translation options.***
- ***Tap, hold and release on a noun and the X-Ray box appears.***
- ***Drag your finger and select a word or phrase to***
    - ***Highlight it.***
    - ***Type a Note about it.***
    - ***Share it.***
    - ***Search for it throughout the book you're reading, every***

book in your Library, or the kindle store.

- ***Tap a highlighted passage to Delete the highlight.***
- ***Tap the top of the screen to display the Standard toolbar. Then tap the [3 vertical dot] icon to launch the Vocabulary Builder. You use that to test your knowledge of words that you've looked up in your Paperwhite's dictionary.***

6.2 Dictionary

Book > tap, hold and release on a word > Dictionary

In a book or PDF document, tap, hold and release on a word and a box will appear with its definition. I use this feature all the time. It's amazing. For most words, information like part-of-speech, pronunciation, and usage examples are also included.

Tap the dictionary name at the bottom right corner of the box and another box appears. From here you can choose to use a different installed dictionary. My Paperwhite has the Oxford dictionary of English and the new Oxford American dictionary installed. You can install any number of the 50+ other dictionaries from your amazon account in a web browser (chapter 10).

Tap outside the box to close it and return to your book.

6.3 Wikipedia

Book > tap, hold and release on a word or phrase > swipe > Wikipedia

In a book, tap, hold and release on a word or phrase, and the dictionary box will appear. Choose Wikipedia from the dropdown list in the top left corner of the dictionary box, or swipe left.

Tap outside the box to close it and return to your book.

6.4 Translation

Your Paperwhite can translate between many languages. According to threads on forums, its with varying levels of success. It's a feature that's bound to improve over time. You can translate a word or phrase, or a passage of text.

6.4.1 Translate a Word or Phrase

Book > tap and hold on a word or phrase > swipe > swipe > Translation

In a book, tap, hold and release on a word or phrase, and the Dictionary box will appear. Choose Translation from the drop-down list in the top left corner of the dictionary box, or swipe left twice. Select the From and To languages from the drop-down lists, and the translation will be displayed.

Tap outside the box to close it and return to your book. The passage will be highlighted. Tap the highlight to delete it if you want.

6.4.2 Translate a Passage

Book > tap and drag on a passage > [3 vertical dots] > Translation

In a book, tap, hold and release on a passage. Tap the [3 vertical dot] icon in the toolbar that appears, and then Translation. Select the *From* and *To* languages from the drop-down lists and the translation will be displayed.

Tap outside the box to close it and return to your book.

6.5 X-Ray

X-Ray lets you find extra information about the characters, places, and terms mentioned in a book. You can use X-Ray to see the first mention of a selection. Or you can use X-Ray to see every mention of a selection in your book.

Not all books have this functionality. But when they do, it's magic. Especially if your memory for names and places is as bad as mine! Check for X-Ray enabled on the book's sales page. Otherwise, tap the [3 vertical dots] icon in the Standard Toolbar (label (36) in Figure 5.4). The X-Ray menu item will be greyed out if it's unavailable.

X-Ray is a really useful tool. Armed with the brief overview I'm about to give you, I encourage you to take the time to explore it.

6.5.1 X-Ray - Single items

Book > Tap and hold on a word

Tap and hold to select a place name or character name and the X-Ray box will appear. It shows you the sentence containing the first time the selection is mentioned in your book.

6.5.2 X-Ray - All items

Book > tap and hold on a word > Open X-Ray

Tap and hold a place name or character name to select it and the X-Ray box will appear. Tap Open X-Ray in the bottom right corner of the box to launch another box that has your selection as its title. The box has a notable clips tab and an all mentions tab.

- **The Notable clips tab:** The Notable Clips tab is where you can view passages relating to your selection that other readers have highlighted. Notable Clips are common in popular books. The tab has a bar with a marker underneath indicating your current place in the book. That's so you can avoid looking ahead and finding spoilers. There's scrolling functionality for long clips, and small (hard-to-see) next and previous mention arrows.

- **The All mentions tab:** The All Mentions tab is where you can see every mention of your selection in the book, and a tappable link to each mention.

6.5.3 X-Ray – From the [3 vertical dots] icon

Book > [3 vertical dots] > X-Ray

While reading a book, tap the top of your screen to show the Standard toolbar, then tap the [3 vertical dots] icon (label (36) in Figure 5.4). From this menu, tap X-Ray, and in the menu that appears, you'll see tabs for Notable Clips, People, Terms, and Images. In each tab, items are noted by name, and the number of times that they're mentioned. The results can be sorted according to their relevance, or alphabetically. Tap any item to be taken to that place in your book.

6.6 Highlighting passages

Book > tap, hold and release on a word or passage > highlight

While reading a book, tap, hold and release on a word or passage. An Options box will launch. Tap outside the box and the selection will be highlighted.

Any passages you highlight get saved to your Amazon account in the cloud. You can access them from the Notes and Highlights menu (label (33) in Figure 5.4). If it's a Kindle Unlimited or Prime Reading book that you've returned, your highlights will be reinstated if you borrow the book again.

You can also view your Highlights in the kindle cloud reader (chapter 10), without having to borrow the book again.

You can also color code your highlights in the kindle cloud reader, and in the kindle apps on your desktop or smart device.

6.7 Adding sticky notes

Book > tap, hold and release on a word or passage > Note

You add a note using the on-screen keyboard. When you add a note, a superscript symbol appears after the text you highlighted. On the Paperwhite these are called sticky-notes. The October 2024 release of the Scribe also has an active canvas tool for on-screen notes.

To add a note: Tap and hold to highlight the text you want the note to be about. Use the keyboard that appears to write your note.

To view your notes: Tap the Notes icon (label (33) in Figure 5.4), or tap the note superscript that's immediately after the highlighted text.

To delete a note: Tap the note's superscript and then the Delete button in the dialog.

Important features:

- All notes are automatically saved and synced to your free Amazon kindle cloud account.

- Amazon reinstates your notes if you reborrow a Kindle Unlimited or Prime Reading book.

- You can access all notes, even for returned books, at rea

d.amazon.com (chapter 10).

- The Notes and Highlights menu (label (33) in Figure 5.4) shows pages with sticky notes.

I use this feature all the time. For example, in the final review of this book, there's only a handful of pages with notes. Rather than scrolling through every page to find my markups, I tap the Notes and Highlights icon and I'm shown only the relevant pages.

6.7.1 Share notes and highlights

Book > tap on a highlighted word or passage > . . .shares to email or goodreads

OR

Book > tap top for Standard toolbar > Notes & Highlights (label (33) in Figure 5.4) > . . .shares to email only.

Because each Kindle device has its own email address, you can share notes and passages that you highlighted. Your Paperwhite sends text that you've highlighted along with a reference to the book. Here's an unedited example of a highlighted passage that I shared to my personal email.

> "Stacey pushed aside any rule she didn't consider worthy of following." (from "The Decoy: A Riveting Espionage Suspense Thriller..." by Chloé Archambault)

Within the sharing box, you can share to an email address or to your goodreads account.

- **Email:** Tap the email button and a prefilled (editable) email box displays. By default, the email includes your highlight as a quote, as well as a link to the book in the amazon store.

Enter an email address using the keyboard that appears, and then tap "send". You can also tick a box to send it to the email address attached to your Amazon account.

- **Goodreads (single highlighted passage only):** Tap the goodreads button and a prefilled (editable) message box displays with your highlight as a quote, as well as a preview of the book. You can add an extra note if you want to. Tap Share to post.

On the Paperwhite, you can't share notes from the note input box. You can only share all notes from the notes and highlights icon in the Standard menu bar (label (33) in Figure 5.4). This emails as a PDF.

6.7.2 Share a book reference

Book > tap top for Standard toolbar > [3 vertical dots] > Share

This sharing option allows you to share a reference to the book you're reading directly to someone's email address or to goodreads. To use the sharing option:

- While reading a book that you've downloaded from the Amazon store, tap the top of the screen to show the Standard toolbar, then tap the [3 vertical dots] icon (label (36) in Figure 5.4). From the menu that appears, tap Share.

- Then select the sharing method. . .

  - **Email:** A prefilled (editable) email box displays. The message contains the full reference for the book, and a link to it in the Amazon store. Enter an email address using the keyboard that appears, and then tap the Send button.

- **Goodreads:** Share the book and your reading progress to your goodreads account. Add a note if you want. Tap the Share button to post.

6.7.3 Export notes

Book > Tap top for Standard toolbar > Notes and Highlights > Share [arrow-in-box icon]

From the Notes and Highlights menu (label (33) in in Figure 5.4), tap the arrow-in-a-box icon to quicksend a PDF of all the notes and highlights from your book to the email address that's attached to your Amazon account. Or share the PDF with up to 5 email addresses. You cannot export individual notes.

This works well for sticky-notes in kindle books, but not so well for sticky notes you've created in a in pdf file. I imagine that's a bug on the paperwhite because it worked okay in a test on my Scribe.

6.8 Vocabulary Builder

Book > Tap top for Standard Toolbar > [3 vertical dots] > Vocabulary Builder

The vocabulary builder records words that you've looked up in the dictionary, and presents them as a grid for you to study and master. It's a great tool to help you remember words, and to improve your vocabulary and reading comprehension. I imagine that it would be incredibly useful to language learners.

You can toggle to show a list of all Words that you've looked up, or according to the Book you looked the word up in. When you tap on a word, a box displays its definition.

There's also an option to show flashcards of random words from the list. Tap the Flashcards label at the bottom of the vocabulary builder to access this. The flashcard displays the word in a sentence from the book you looked it up in, and optionally, its definition too.

Each word is flagged as being one that you're either Learning or have Mastered. There's a 2000 word limit. When you reach that limit you can delete words to make room for more.

Could you spare a moment to review my book?

Thank you so much for reading my book! If you've found it helpful, I'd be incredibly grateful if you could leave a short review. Reviews are a huge help to indie authors. They help other readers discover our books and keep us writing. Just a few words about what you liked would mean the world. Thanks again for your support!

Review on Amazon US	**Review on Amazon UK**
 https://amzn.to/4hFb80w **Click the Flag or Link** **Scan the QR code**	 https://amzn.to/3EA7u9K **Click the Flag or Link** **Scan the QR code**
Review on Amazon Canada	**Review on Amazon Australia**
 https://amzn.to/3Equs3q **Click the Flag or Link** **Scan the QR code**	 https://amzn.to/4hmKEkv **Click the Flag or Link** **Scan the QR code**

Don't forget the resources from the QR codes in the figures, and at ianallanauthor.com/paperwhite

Chapter 7

The Display Settings box

Figure 7.1: The Aa icon launches the Display Settings box (download the figures from bit.ly/akpug or the QR code).

Book > Tap top for the Standard Toolbar > Aa

Your Paperwhite uses a special format called reflowable text. It is similar to text editors like MS Word in that you have lots of ways to customize how a document looks on your screen. That's what this chapter is about - taking advantage of reflowable text to customize the way your books look on your Paperwhite.

While reading a book, tap the top of the screen to reveal the Standard toolbar. Then tap the Aa icon (label (31) in Figure 5.4). The box that appears contains four submenus - Themes (40), Font (41), Layout (42), and More (43). The changes you make in these menus apply to your current book and all subsequent books that you open.

If you're viewing a 'fixed format' document such as a PDF that you sent via sendtokindle (section 8.2.1), then the menu that appears only contains the Layout and More menu items (label (42) and label (43) in Figure 7.1, and section 7.3).

7.1 Themes

Book > Tap top for the Standard Toolbar > Aa > Themes

To display the Themes sub-menu, while reading a book, tap the top of the screen to reveal the Standard toolbar. Then tap the Aa icon, and then Themes (label (40) in Figure 7.1).

Themes are a mix of font size, style, and line spacing that you customize to suit your reading preferences. There are four default settings to experiment with (compact, standard, large and low-vision). In the Manage Themes sub-menu, you can toggle which themes to display. You can create a custom theme by making changes in the four sub-menus (Themes (40), Font (41), Layout (42), and More (43)), and then tapping the Save Current Settings button.

7.2 Fonts

Book > Tap top for the Standard Toolbar > Aa > Font

To display the Font sub-menu, while reading a book, tap the top of the screen to reveal the Standard toolbar. Then tap the Aa icon, and then on Font (label (41) in Figure 7.1).

You can change the font family (there's 10 fonts) and text size to suit your reading preferences, and preview your changes in real-time before applying them. . .

- Use the sliders to adjust the text boldness and text size. There are 5 levels of bold and 14 font sizes.

- Tap on the font name to choose a different font.
- After adjusting the font and text size, tap outside the box to save your changes and return to your book.
- You can also adjust the text size while you're reading a book by using a pinching action with two fingers.

This functionality only applies to books with reflowable text (ePubs, PDFs, and other documents sent to your paperwhite with adjustable text). You cannot change the text size in your Paperwhite's menus or the text in fixed-format PDF documents.

7.3 Layout

Book > Tap top for the Standard Toolbar > Aa > Layout

To display the Layout sub-menu, while reading a book, tap the top of the screen to reveal the Standard toolbar. Then tap the Aa icon, and then Layout (label (42) in Figure 7.1).

You can customize orientation, margins, text alignment, line spacing and columns. . .

- **Orientation:** Choose portrait or landscape.
- **Margins:** Adjust the amount of white space around the edges of the text on the screen.
- **Text alignment:** Choose whether the text is aligned to the left, or justified.
- **Line spacing:** Adjust the amount of space between lines of text on the screen.
- **Columns:** Choose to display as one column or two columns (in landscape orientation).

7.4 More (additional features)

Book > Tap top for the Standard Toolbar > Aa > More

To display the More sub-menu, while reading a book, tap the top of the screen to reveal the Standard toolbar. Then tap the Aa icon, and then on More (label (43) in Figure 7.1).

This menu provides access to additional options for customizing your reading experience. They include Reading Progress, Show clock while reading, Book mentions, About this book, Popular highlights, Highlight menu, Page turn animation and Word Wise.

7.4.1 Reading Progress

Book > Tap top for the Standard Toolbar > Aa > More > Reading Progress

To display the Reading Progress option, while reading a book, tap the top of the screen to reveal the Standard toolbar. Then tap the Aa icon, then on More (label (43) in Figure 7.1), then on Reading Progress.

This option lets you control the text that displays in the bottom left corner of your screen. The options include Page in book, Time left in chapter, Time left in book, Location in book, and None.

While you're reading a book, you can also tap the Reading progress text in the bottom-left corner and advance through the options.

7.4.2 Show clock

Book > Tap top for the Standard Toolbar > Aa > More > Show clock while reading

To display the Show clock while reading option, while reading a book, tap the top of the screen to reveal the Standard toolbar. Then tap the Aa icon, and then on More (label (43) in Figure 7.1), then on Show clock while reading.

The clock displays at the top of the screen while you're reading. This option lets you toggle the clock on and off. You set the clock time in the Settings menu (chapter 9).

7.4.3 Book mentions

Book > Tap top for the Standard Toolbar > Aa > More > Book mentions

To display the Book mentions option, while reading a book, tap the top of the screen to reveal the Standard toolbar. Then tap the Aa icon, and then on More (label (43) in Figure 7.1), then on Book mentions.

Where an author mentions another book, toggling this option on lets you tap on the book mention and be taken to the book in the kindle store. To be honest, I've not had much luck with this.

7.4.4 About this book

Book > Tap top for the Standard Toolbar > Aa > More > About this book

To display the About this book option, while reading a book, tap the top of the screen to reveal the Standard toolbar. Then tap the Aa icon, and then on More (label (43) in Figure 7.1), then on About this book.

When you open a book for the first time, a box appears showing you some general information about the book (eg. popular highlights, more by this author, etc), as well as giving you the option to Mark as currently reading in goodreads. Toggle to enable or disable this functionality.

7.4.5 Popular highlights

Book > Tap top for the Standard Toolbar > Aa > More > Popular highlights

To display the Popular highlights option, while reading a book, tap the top of the screen to reveal the Standard toolbar. Then tap the Aa icon, and then on More (label (43) in Figure 7.1), then on Popular highlights.

Toggle this on to display passages that have been highlighted by other readers. Popular highlights appear with a fine dotted underline, and small text saying eg. "578 highlighters". The useful thing about this functionality is that, well, maybe a passage that's been highlighted by 578 people is one that you should pay greater attention to? Also, if you write in the same genre you read, it's a useful market research into themes that you should consider emulating.

7.4.6 Highlights menu

Book > Tap top for the Standard Toolbar > Aa > More > Highlights menu

To display the Highlights menu option, while reading a book, tap the top of the screen to reveal the Standard toolbar. Then tap the Aa icon, and then on More (label (43) in Figure 7.1), then on Highlights menu.

Toggle this on and off. This menu is important if you like to take notes. When you highlight a piece of text. . .

- If the highlights menu is toggled on, a menu appears asking you if you want to highlight it and make a note about it.

- If the highlights menu is toggled off, the text is highlighted, and no note taking menu appears.

7.4.7 Page-Turn Animation

Toggle on to animate your page turns.

To display the Page-Turn Animation option, while reading a book, tap the top of the screen to reveal the Standard toolbar. Then tap the Aa icon, and then on More (label (43) in Figure 7.1), then toggle Page-Turn Animation on.

The animation looks a bit like a swipe transition in PowerPoint.

7.4.8 Word Wise

Book > Tap top of screen for the Standard toolbar > Aa > More > Word Wise

To display the Word Wise option, while reading a book, tap the top of the screen to reveal the Standard toolbar. Then tap the Aa icon, and then on More (label (43) in Figure 7.1), then on Word Wise.

This menu option lets you toggle Word Wise on and off, and adjust the level of help it provides.

Word Wise provides short above-text definitions for challenging words. The lines in your book become double-spaced to make room for the definitions to display in a tiny font above the word. You can also tap on a word that Word Wise recognizes and a box will launch with a more detailed definition.

Wordwise seems to only work in Kindle books. At least, not for the PDF files in my library. If you tap on Word Wise in the bottom right of your screen, that launches a slider that gives you the ability to show more or fewer hints.

I find Word Wise to be distracting, but I imagine that it would be incredibly useful to language learners.

Chapter 8

Buying, transferring and managing books

In this chapter I'll show you how to add books to your Paperwhite from its kindle store, and sources such as the kindle app, the amazon app, and the amazon website. I'll also show you how to transfer books, documents and images from your computer using drag-and-drop from a web browser, as an email attachment, and directly using the cable that came with your Paperwhite. Finally, I'll show you how to send a web page and an MS Word document to your Paperwhite, and how to organize your content into collections.

8.1 Quickstart

- ***Tap on a thumbnail in the Home screen to be taken to a book's sales page.***
- ***Tap on the shopping cart icon in the Home screen to be taken to the kindle store.***
- ***Tap on the Buy button on a sales page to buy a kindle book.***
- ***Kindle Unlimited subscribers and Prime members can tap***

the Read Now button to borrow an eligible book.

- ***Add a non-amazon book to your Paperwhite using the drag-and-drop amazon.com/sendtokindle web page.***
- ***Install the Send-to-Kindle app and then drag-and-drop content into it.***
- ***Share a Word document to your Paperwhite directly from MS Word.***
- ***Share a webpage to your Paperwhite using Share on your smartphone, or the Send To Kindle chrome extension on your desktop.***

8.2 Transferring content

You can transfer a book, document or image from your computer to your Paperwhite using the sendtokindle web page, the send-tokindle app, as an email attachment, or directly via a cable. You can also send a Word document directly from MS Word. The simplest and most reliable method is the amazon.com/SendToKindle drag-and-drop web page or app, followed by sending a file as an email attachment.

SendtoKindle and emailing your Paperwhite with an attachment overcome a bunch of the problems that you're likely to have if you use cable transfer. Sometimes your Paperwhite won't recognize a cable-transferred file but will recognize the same file if it's sent via the Amazon cloud. That's because Amazon checks your file for compatibility, rejects incompatible file types, sometimes mends incompatible files, and places the book or file onto your Paperwhite using the correct folder structure standards.

You can transfer twelve file types to your Paperwhite - .azw, .epub, .mobi, .PDF, .doc, .docx, .rtf, and .html document files. Also .png, .jpg, .gif and .bmp image files.

You can also send a web page to your Paperwhite using Send on your smartphone and tablet, or the Send To Kindle extension for google chrome. I sometimes read longform newspaper articles that way.

Now let's look more closely at these computer-to-Paperwhite file transfer techniques.

8.2.1 SendToKindle drag-and-drop transfer

Amazon.com/sendtokindle

OR

Drag-and-drop to one of the Kindle Apps (iOS, Windows, Mac, see Figure 8.1)

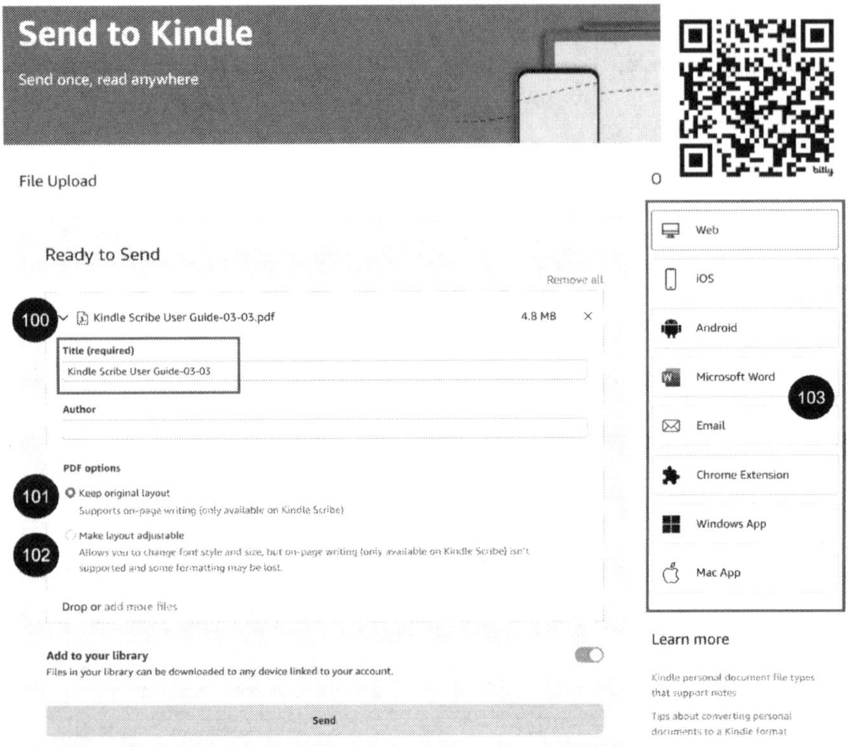

Figure 8.1: The amazon.com/sendtokindle dialog box (download the figures from bit.ly/akpug or the QR code).

Supported File Types are PDF, DOC, DOCX, TXT, RTF, HTM, HTML, PNG, GIF, JPG, JPEG, BMP, EPUB.

Sendtokindle sends some formats (eg. .docx) only with an adjustable layout. Those files appear on the paperwhite like a kindle book, and you can't mark them up. In MS Word, you can export directly from the "File -> Export -> Send documents to kindle" path. This has a fixed format option.

Here's a detailed rundown of how to use sendtokindle with a PDF file.

- Login to your Amazon account.

- Go to amazon.com/sendtokindle.

- Click the arrow to the left of the title to see the export options (label 100 in Figure 8.1).
- Choose the layout option that best meets your annotation needs.
  - **YES:** Keep original layout for full writing and highlighting capabilities (label (101)). On the Kindle Scribe, your document can be drawn on, written on, highlighted, and marked up as you would with a pen and highlighter on paper.
  - **NO:** Make layout adjustable for reflowable text (label (102)). Your document will behave like an eBook (ie. adjustable text size, and sticky notes). If you have figures or tables in your document, this technique can turn the document into gobbledygook!

After you hit send, the dialog will show your PDF as processing, and then in library, and ready to open and markup on your Scribe. Be patient if search isn't working on a freshly downloaded PDF. Sometimes it can take a little while to be indexed.

8.2.2 Email transfer

[3 vertical dots] > Settings > Your Account > Send-to-Kindle email

OR

Amazon in a web browser > Accounts & Lists drop-down menu > Manage content and devices > Devices > Amazon devices > (now pick your Paperwhite from the list and note its email address in the Device Summary Area)

OR

YourCountry'sAmazonStore/myk > Devices > Amazon devices > (now pick your Paperwhite from the list and note its email address in the Device Summary Area)

To email a book from your computer to your Paperwhite . . .

You will need to send it to your Paperwhite's email address: Here's where you find that. . .

- **To find the email address on your Paperwhite:** Swipe down from the top of the screen for the Quick Actions toolbar, then tap All Settings, then Your Account and then Send-to-Kindle email.

- **To find the email address in a web browser:**

    - From your local amazon web page, click Your account, then Manage content and devices, then Devices, then Amazon devices, then Amazon devices (and pick your Paperwhite from the list).

    - From your local amazon web page, add **/myk** to the URL (eg amazon.com/myk amazon.co.uk/myk). Click Devices, then Amazon devices, and then pick your Paperwhite from the list.

Pay attention when you're copying the email address. Unfortunately, kindle email addresses often contain both the number zero and the letter O, and they look near-identical. If you look closely (I zoom in with my smartphone camera), the letter O is more circular and the number zero is more egg shaped (O the letter versus 0 the number).

- **You need to send the file from an approved email address:** For security reasons, the email address you're sending from must be on your approved personal document email list. I talk about that later in this chapter.

- **You can only send supported file types:** Only file types that are supported by Kindle will be delivered to your device (.azw, .epub, .mobi, .PDF, .doc, .docx, .rtf, and .html document files. Also .png, .jpg, .gif and .bmp image files).

With those three things in place (ie. your Paperwhite's email address, an approved from-email address, and compatible file type), you can now send a file as an attachment to your Paperwhite's email address. The file will be checked and processed by Amazon, and then delivered to your Paperwhite. Be patient. Most times you'll only need to wait a minute or two, but I have waited up to 15 minutes a few times.

PDF files keep their format when you email them to your Paperwhite. However, if you email a PDF file with the word "convert" in the subject line, amazon will convert it into reflowable text (reflowable text) as a part of the transfer. Be aware that "convert" can sometimes reorganize and jumble your PDF.

8.2.3 Cable transfer

Transfer PDF files from your computer to your Paperwhite using the USB cable that came with it (see also chapter 12). This technique no longer works reliably. I spent a frustrating hour experimenting and could only get it to work for PDF files. Don't use it. You'll just end up angry and swearing. However, if you insist . . .

1. Connect the cable to your computer, and your Paperwhite will appear in your computer's file manager as a removable disk. You can drag-and-drop, or copy-and-paste your files to your Paperwhite as you would to any external disk.

2. Copy the PDF file from your computer to your Paperwhite's documents folder.

3. After transferring a book, eject your Paperwhite from your computer and the PDF will appear in your Library.

8.2.4 Send web page via Chrome.

On your desktop, in the Chrome browser, google "send to kindle chrome extension". Click on the link to the Chrome store and then the add to desktop button. Once the extension is installed. . .

- Login to your amazon account.
- In a webpage, tap the Kindle button that the extension added to your browser toolbar. That will open the Send To Kindle dialog box.
- In the settings menu, toggle "Add to your library" button...
    - **On** to add the webpage available to all your kindles (devices and apps), or
    - **Off** to choose from the list, the kindle device that you want to send the webpage to.
- Choose either *Quick send* or *Preview and send.*

8.2.5 Send web page from smart phone.

You can send a web page to your Paperwhite from your smartphone or tablet.

- Add your email to your Paperwhite's approved email address list (see below).
- In a web page, tap the share button and Kindle as the app

to share with. You may also need to have the kindle app installed on your device and connected to your account for this share option to appear. If you get an error then try using a different browser.

8.2.6 Manage your Paperwhite's email address

Amazon in a web browser > Accounts & Lists drop-down menu > Manage your content and devices > Preferences > Personal document settings > [then view your Paperwhite's details]

OR

Amazon in a web browser > Accounts & Lists drop-down menu > Your account > Content and devices > Preferences > Personal document settings > [then view your Paperwhite's details]

OR

YourCountry'sAmazonStore/myk > devices > amazon devices > [then view your Paperwhite's details]

You can change your Paperwhite's email address if you don't like the one that amazon assigned to it. Also, if you plan to email a file to your Paperwhite, you'll need to add the address that you'll be emailing from to your Paperwhite's approved email-from-addresses. You manage these email addresses in the Personal document settings area of your amazon account. . .

- **To change your Paperwhite's email address:** Every one of your kindle devices (Kindle apps and other kindle devices) has its own email address. Click the edit button next to the Send-to-Kindle email to change your Paperwhite's email address. You should make it a hard-for-spammers-to-guess email address. Otherwise you may find that amazon starts to send you validation emails every time you, or a spammer, sends an email to your Paperwhite.

- **Approved Personal Document E-mail List:** Your Paperwhite can only accept emails (and attachments) from addresses that are listed on your account as approved send-from email addresses. Click the Add a new approved e-mail address link in the Approved Personal Document E-mail List area and add the email address that you'll be sending your files from.

8.3 Getting books from the Kindle store

Tap on a book thumbnail on your Paperwhite, in the kindle app, or in the amazon store in a web browser and you'll be taken to its sales page. There you can download a sample of the book, buy it, or borrow it if it's an eligible book and you're a Prime member or Kindle Unlimited subscriber.

8.3.1 Sampling books

On a book's sales page, tap Try a sample to download. The sample (usually 10% of the book) will show up in your Library. Tap the thumbnail to read it, and on the sample's last page Amazon will give you the option to buy it.

You can delete a sample by tapping and holding on the book thumbnail and then tapping on Remove this sample in the menu that appears.

8.3.2 Buy books from your Paperwhite

To buy a book, tap on its thumbnail in the Home screen or from the kindle store that launches from the Shopping Cart icon (label (3) in Figure 5 1), and you'll be taken to the book's sales page in the kindle store. When you tap the Buy button on the book's sales page, the book will automatically download to your Paperwhite and be processed using the payment method associated with your Amazon account.

8.3.3 Buying books from a web browser

Sign in to Amazon on your browser. From the drop down list. To the left of the search bar, choose Kindle Store, and then tap the magnifying glass to its right. You'll be presented with the store's best sellers and a bunch of drop down lists on the top menu bar. Explore! Categories is probably the best drop-down list to start with.

Tap on a book's thumbnail and if you decide to buy it, tap the button that sometimes reads Buy now with 1 click. For Prime members and Kindle Unlimited subscribers, this may say read for $0.00. The book will be delivered to your Paperwhite, and to every kindle device that's linked to your Amazon account. I'm in Australia, and I've found that sometimes books that are unavailable on amazon.com become available when I check my local store, and visa versa.

8.3.4 Borrow books from amazon

You can borrow a book from the kindle store if you're a Prime member or a Kindle Unlimited subscriber (chapter 11). When you're logged in, Amazon automatically applies your membership and subscriptions to the buy button on a book's sales page, and will show the correct price for your membership level. Kindle unlimited subscribers can borrow any book labelled Kindle Unlimited, and Prime members can borrow a constantly changing subset of Kindle Unlimited books.

8.3.4.1 Borrow books from your Paperwhite

Tap the button, sometimes labelled as Read now and sometimes labelled as Buy for $0.00, and the book will download. Books are labelled as Kindle Unlimited if they're enrolled in the scheme. You can make the store show only those books by using the Prime and Kindle Unlimited category filters at the top of the kindle store (label (4) in Figure 5 1).

8.3.4.2 Borrow books from a web browser

Login to the Amazon website and find the book you want to borrow. Tap the Read for Free button and it will be delivered to your amazon cloud and by default to all your kindle devices.

Over broadband internet, your book should download and appear on your home screen in less than a minute. Tap the thumbnail to start reading.

8.3.4.3 Borrowing books from the amazon app

If you're a Prime member or Kindle Unlimited subscriber, you can borrow books from the Amazon app. You cannot buy Kindle books from the amazon app. That's because app stores charge sellers a fee, and Amazon doesn't want to pay it!

Once you've found a book, from the Deliver to tab, select your Paperwhite as the device you want the book to download to, and then click Read for $0.00. Amazon's Whispersync will deliver the book directly to your Paperwhite. Be patient. It can sometimes take a minute or two to arrive (not 15 minutes).

8.3.4.4 Borrowing books from the kindle app

If you're a Prime member or Kindle Unlimited subscriber, you can borrow books from the kindle app.

Once you've found a book, click Read for $0.00. Amazon's Whispersync will deliver the book directly to your kindle cloud – that's all your kindle devices including your Paperwhite. I've found that sometimes a book won't download to my Paperwhite unless I've opened it in the app first.

You can't buy Kindle books from the kindle app. That's because app stores charge a seller's fee (that amazon doesn't want to pay).

8.3.5 Borrowing books from a public library

Kindle books can only be borrowed from US libraries that subscribe to Overdrive. You need to be logged in to both your local library and your amazon account.

- You can send an eBook to your Kindle via Overdrive's Libby app. On the library's book page, check to see if Kindle is one of the supported formats. Click the link and follow the prompts.

- You can also borrow a Kindle book via Overdrive and read it on the amazon cloud reader.

For more information. . .

- Search google for the following title "Borrowing Kindle Books from your library's OverDrive website or the OverDrive app" help.overdrive.com/en-us/0431.html

- And Search google for this title too "libbyapp reading with kindle" help.libbyapp.com/en-us/6017.htm#

The 9th Street Books website has compiled a list of libraries that cater for non-residents. Google "library access for everyone 9th street books" to find it. 9thstreetbooks.com/how-to-get-a-library-card-online/#3-libraries-offering-library-cards-online-for-non-residents

Also, recently I stumbled on a video that explained how non-residents could borrow from the Chicago public library. To find the video, google search "borrow a kindle book from Chicago library".

8.3.6 Finding free books

You don't have to be an Amazon Prime member or a Kindle Unlimited subscriber to get free Kindle books. You could simply choose to only download free books from the Amazon store! Often authors list their books for free in the hope that they will get reviews, or the book might be first in a series that they hope to get you hooked on.

Here's some places to look for free books. . .

- In the Kindle store search bar (label (1) in Figure 5 1), type terms like "free Kindle books" and "Classic literature".
- OverDrive and Libby in participating US libraries
- gutenberg.org
- standardebooks.org
- manybooks.net
- search google for "free Kindle books"
- search Facebook and other social media for "free Kindle books"

- Signup to Bookbub.com. They send daily emails with links to free and low cost books (usually 99 cents)
- Signup to sites like booksirens.com, booksprout.co if you're up for reviewing pre-release books (often by big-name authors). These sites will keep sending you free books so long as you keep reviewing them.

If you find a book on a non Amazon site, you're best to download it in ePub format. Then transfer it to your Paperwhite using one of the techniques I described earlier in this chapter.

8.4 Removing books

In this section I show you how to return a Kindle Unlimited or Prime borrow, return an unwanted or accidental purchase, and how to delete or remove a book from your Paperwhite. . .

8.4.1 Return borrowed books to Amazon

Home or Library > tap and hold the book thumbnail > Return to Kindle Unlimited

To return a Kindle Unlimited or Prime reading book, in your Library screen, tap and hold your finger on the book thumbnail to open the book menu. Then tap on Return to kindle Unlimited. This will remove the book from your library and end your access to it. Your notes and highlights are saved and will be reinstated if you borrow the book again. You can access your notes and highlights any time from read.amazon.com without having to reborrow the book (chapter 10.2).

To show only the Kindle Unlimited and Prime Reading books in your Library, tap the filter button (label (4) in Figure 5 1) and choose Kindle unlimited from the Your Programs area in the box.

If you need to see your Kindle Unlimited borrowing history, look in the Content and Devices area of your amazon account (chapter 10).

8.4.2 Return a book for a refund

Amazon in a web browser > Returns & Orders > Return for a Refund

There's three ways to return a book for a refund.

- **If you purchased it by accident:** There's a Purchased by accident? area on the Thankyou Page that displays following a successful purchase. Tap the CANCEL ORDER button to return the book and get a refund.

- **If it is within 7 days of buying the book:** In a web browser, go to Your Orders in your amazon account. Then click Return for a Refund and from the drop-down list, choose a reason for the refund (eg. wrong item, defective, etc).

- **If you have read more than 10% of the book:** You will need to send a customer service request. This will be reviewed by a human to ensure your return complies with Amazon's abuse policies.

8.4.3 Delete books from your library

Even though your Paperwhite stores a lot of books, you might decide to free space or to tidy up.

You can choose to remove download, or permanently delete a non-kindle-unlimited book or file that you've downloaded to your

Paperwhite. Tap and hold on a book's thumbnail, and from the menu that appears. . .

- **Remove download:** tap to keep the book in the cloud but remove it from your Paperwhite.
- **Permanently delete:** tap to remove it from your Paperwhite and the cloud.

You can also delete a book from within any of your kindle apps, or from your amazon account in a web browser.

8.5 Using Collections

Collections are a bit like folders on your desktop computer and can be a great way to organize your books. For example, you might want to create a collection called fiction and another called nonfiction.

You can create a collection on your Kindle Paperwhite, any other Kindle device, or the Kindle app on your phone. Once created, the collection will sync across all your Kindle devices. Let's look at how to create, add and view them.

8.5.1 Creating a collection

Library > + > Create a Collection

To create a collection, from the Library screen, tap + in the top right corner (label (5) in Figure 5 1) and the Create a collection dialog will appear. Use the keypad that appears to give your collection a name.

There is a bug in the Paperwhite's menu system. If the + icon is not there, tap the sort order icon (label (6) in Figure 5 1) and then

Collections in the View options area. That should force the + icon to appear.

8.5.2 Adding books to a collection

Library > tap and hold a book thumbnail > Add to / Remove from collection

To add a book to a collection, in your library, tap and hold the book, and then from the popup menu, choose add to / remove from a collection and then tap the collection you want to add it to.

8.5.3 Viewing a collection

Library > Filter > Collections

You can access the collection from your Library screen by selecting the Collections filter from the three horizontal bar menu (label (6) in Figure 5 1). Dismiss the filter box by tapping outside it. Tap on the collection thumbnail to see the books within it.

Chapter 9

The Settings menu

The Settings menu is a collection of options and settings that you can use to customize and manage your Paperwhite. The Settings menu can be accessed from the. . .

- Quick access toolbar within a book (label (21) in Figure 5.3).
- Standard toolbar (label (36) in Figure 5.4).

To simplify the remainder of this chapter, I'm going to assume that you've already found your way to each of the Settings sub-menus via one of those paths.

9.1 Your Account

[3 vertical dots] > Settings > Your account

From the Your Account sub-menu, you can access parental controls, deregister your Paperwhite, change your device name, add personal information, connect to Goodreads, and find your Kindle email address.

9.1.1 Parental Control

[3 vertical dots] > Settings > Your account > Parental Control

The Parental Control sub-menu lets you to supervise your child's access to books and kindle functionality on your Paperwhite. Amazon kids, where it's available, lets you manually add and remove books or automatically give your child access to age-appropriate content in the store. The restrictions area acts as a general security item at the device level.

9.1.1.1 Amazon Kids

[3 vertical dots] > Settings > Parental Controls > Amazon Kids

Unfortunately, amazon kids is only available in some countries (not where I live). Where it is available it appears as a separate Amazon Kids item in the Parental Control area. From there you can add a child's name and date of birth. This lets you automatically restrict some content according to age appropriateness. Alternatively, you may choose to give them access to individual books.

You can also manage your child's content from the digital content area of your amazon account (amazon.com/myk in a browser). From the more actions menu next to a book, take the Manage family library option and add the book to an Amazon Kid's Library.

This appears to be a well-thought-out feature that flows through into other Amazon subscription products too. I'm disappointed that I can't give you a better overview. Following are links to two YouTube channels. A search for Parental controls in each channel will produce a half dozen or so useful short videos.

- **@HardResetinfo:** youtube.com/@HardresetInfo
- **goodereader:** youtube.com/goodereader

9.1.1.2 Restrictions

[3 vertical dots] > Settings > Parental Controls > Restrictions

Tap Restrictions to limit the functionality of your Paperwhite.

In those countries where Amazon Kids is not available, you can set controls at the device level. That's great, not just for kids, but also for when you lend your Paperwhite to another adult, but say, are happy for them to access all the Paperwhite's functionality, except the ability to buy a book. You can toggle the options on and off, but you'll need to enter your parental control PIN to do that. Here's what the options do. . .

- **Web browser:** Removes access to the web browser from the [3 vertical dots] menu, and from any links within a book.

- **Store:** Removes access to the store so that books cannot be bought or borrowed.

- **Cloud:** Users will only be able to access content that is stored on your Paperwhite.

- **Goodreads:** Removes access to the goodreads site, so you cannot add currently reading status or review a book.

9.1.1.3 Parental control PIN

[3 vertical dots] > Settings > Parental Control > Parental control PIN

Tap Parental control PIN and you will be prompted to enter your current PIN. After you have entered your current PIN, you will be able to enter a Parental control PIN of your choice.

Enter the new PIN and then again in the confirmation box.

9.1.2 Deregister device

[3 vertical dots] > Settings > Your Account > Deregister Device

OR

Amazon in a web browser > Accounts & Lists drop-down menu > Manage your content and devices > Devices (select device) > Deregister

OR

Amazon in a web browser > Accounts & Lists drop-down menu > Manage your content and devices > Devices (select device) > Deregister

OR

YourCountry'sAmazonStore/myk > Devices > Amazon devices > pick your Paperwhite from the list > Deregister

To deregister your Paperwhite, tap Deregister Device. Then tap Deregister to confirm that you want to deregister the device.

You deregister your Paperwhite if its been lost or stolen, or if you need to transfer its ownership to another person or another account. Deregistering removes your personal information and settings, and unlinks your Paperwhite from your amazon account. Afterwards, you can setup your Paperwhite again and register it to a different Amazon account (chapter 3 and chapter 4).

9.1.3 Personal Info

[3 vertical dots] > Settings > Your Account > Personal Info

Tap Personal Info to add personal information about yourself. It's a free text box that could include your name, email address, phone number, etc.

9.1.4 Goodreads

[3 vertical dots] > Settings > Your Account > Goodreads

Tap Goodreads to connect to the 125 million strong goodreads book reviewing site.

When you're linked to goodreads, you can share that you're currently reading a book, your reading progress and your star rating for the book. Select the Link your account option and follow the on-screen instructions to sign in to your goodreads account.

You Paperwhite's connection to goodreads is enhanced by all the extra things you can do when you're logged into the goodreads site in a web browser. In the browser version, you can connect with other readers, access interest focused lists, participate in competitions, and leave written reviews.

9.1.5 Send-to-Kindle email

[3 vertical dots] > Settings > Your Account > Send-to-Kindle email

The Send-to-Kindle email is an email address that is associated with each of your Kindle devices and each of your installations of the kindle app. You can use this email address to send documents, books, and other files from your computer or other devices directly to your Paperwhite without the need for a cable or other physical connection.

You can also find this address in the devices area of your amazon account in a web browser. The email address that you're sending from must be in the approved send-from email list (chapter 8).

9.2 Wi-Fi settings

[3 vertical dots] > Settings > Wi-Fi

The wireless sub-menu includes options for placing your Paperwhite into airplane mode, connecting to a Wi-Fi network, and deleting Wi-Fi passwords.

9.2.1 Airplane mode

[3 vertical dots] > Settings > Wi-Fi > Airplane Mode

When airplane mode is toggled on, your Paperwhite's Wi-Fi is turned off and it disconnects from all Wi-Fi. It's essential functionality for air travel. Airplane mode can also help to conserve battery life.

Airplane mode is also available from the Quick Actions toolbar (label (18) in Figure 5.3).

9.2.2 Wi-Fi networks

[3 vertical dots] > Settings > Wi-Fi > Wi-Fi networks

Tap Wi-Fi networks to be shown a list of available Wi-Fi networks. Tap a network name to connect to it. Your Paperwhite remembers networks that you've connected to before. I troubleshoot Wi-Fi connection problems in chapter 12.

When you're connected to Wi-Fi, you're connected to the Amazon ecosystem - the kindle store, synchronized reading across kindle devices, device management, and more.

9.2.3 Delete Wi-Fi Passwords

[3 vertical dots] > Settings > Wireless > Delete Wi-Fi Passwords

Tap Delete Wi-Fi Passwords to permanently delete ALL Wi-Fi passwords stored on your Paperwhite.

9.3 Device options

[3 vertical dots] > Settings > Device Options

The Device options sub-menu lets view information about your Paperwhite, check for software updates, set a date and time, manage languages and dictionaries, manage privacy, free up storage space, manage Whispersync, set your Paperwhite's sleep regime, power saving, restart it, and reset it.

9.3.1 Device info

[3 vertical dots] > Settings > Device Options > Device Info

Tap Device Info to view a collection of information about your Paperwhite, including its name, model number, software version, serial number, and other details that can be useful if you need to check hardware and software versions when you're troubleshooting a problem.

You can also change the name of your Paperwhite in here. I changed mine to "Ian's Paperwhite".

9.3.1.1 Device name

[3 vertical dots] > Settings > Device options > Device info > Device Name

Tap Device name to give your Paperwhite a name. Sometimes you need to send books to an individual kindle, so make sure it's a name you'll recognize. For example, I have Kindle devices called Ian's Scribe and Ian's Paperwhite. My Kindle app installations are called Ian's iPad, Ian's iPhone and Ian's desktop.

You can also change the name of your Paperwhite (or any of your Kindle devices or installations) from the content and devices area of your amazon account.

9.3.1.2 Additional technical information

[3 vertical dots] > Settings > Device options > Device info

This area includes information such as your Paperwhite's network address, its serial number, firmware version and available disk space.

9.3.2 Software updates

[3 vertical dots] > Settings > Device options > Device info > Software updates

- Tap to update to the latest Paperwhite software version. Its wise to check this if your Paperwhite is misbehaving.
- Tap Update Your Kindle to update your Paperwhite.

Occasionally, Amazon releases software updates to add new features and fix bugs. Your Paperwhite should check for these automatically so you're unlikely to use this button. If no updates are available, this option is greyed out.

To update your Paperwhite, you need to be connected to Wi-Fi. It's bad practice to allow any device to shut down part way through an update, so make sure it has enough charge.

You can also install updates manually. Search google for "Kindle E-Reader Software Updates" to find the update page and instructions on how to install it.

9.3.3 Date and Time

[3 vertical dots] > Settings > Device Options > Date and time

Tap Date and Time to set the time. Tap the up and down arrows for hours, minutes and am/pm.

Toggle to choose a 12 hour of 24 hour clock. There's also an option to set the time automatically from the Amazon cloud. Amazon knows where you live because you have a delivery address attached to your account.

9.3.4 Languages and dictionaries

[3 vertical dots] > Settings > Device options > Languages and Dictionaries

Tap the Languages and Dictionaries sub-menu to manage the language and dictionary settings on your Paperwhite. The menus include options for changing the device's language, changing dictionaries, changing the keyboard layout, and Chinese sort order.

9.3.4.1 Language

[3 vertical dots] > Settings > Device options > Languages and Dictionaries > Language

Tap Language to choose the language that corresponds to the book that you're reading. If you're reading a book written in English you have a choice between US English and UK English. If you're reading a book written in French, you would choose French or Canadian French, etc. Incompatible language selection is a major reason why some books display crazy characters.

9.3.4.2 Dictionaries

[3 vertical dots] > Settings > Device options > Languages and Dictionaries > Dictionaries

Tap Dictionaries to choose the dictionary you want to use. In the English language, you have the choice between the Oxford Dictionary of English and the New Oxford American Dictionary.

Not every one of the 50+ other dictionaries is available on your Paperwhite by default. You install additional dictionaries from your web browser (chapter 10).

9.3.4.3 Keyboards

[3 vertical dots] > Settings > Device options > Languages and Dictionaries > Keyboards

Tap Keyboards to change your Paperwhite's keyboard to any one of 27 different language preferences.

9.3.4.4 Chinese Sort Order

[3 vertical dots] > Settings > Languages and Dictionaries > Chinese Sort Order

Tap Chinese Sort Order to change the way that Chinese characters are displayed. You can toggle between Hanyu Pinyin, Stroke order and Zhuyin Fuhao.

9.3.5 Privacy and security

[3 vertical dots] > Settings > Device options > Privacy and Security

The options in this area allow you to set a PIN for your Paperwhite, and choose whether or not to share information about your reading habits with Amazon.

9.3.5.1 Device PIN

[3 vertical dots] > Settings > Privacy and security > Device PIN

Tap Device PIN to set a Personal Identification Number (PIN). Without a PIN, anyone who picks up your Paperwhite can buy and review books on your amazon account, as well as reading or deleting your books and personal documents.

To set a passcode, enter a PIN, and then re-enter to confirm.

Once you've set a passcode, you'll be prompted to enter it whenever you turn on the device or wake it from sleep mode.

Be security conscious. Don't use the same PIN as you use for your banking!

Here's a way to choose a PIN that you won't forget. Think of an obscure town that you visited once, or a town your grandparents lived. Use its ZIP code. Then, if you forget your PIN, hopefully a town's name and your association with it can act as a memory trigger.

In chapter 12, I show you what to do if you forget your PIN

9.3.5.2 Privacy

[3 vertical dots] > Settings > Device Options > Privacy and security > Privacy

Tap Privacy to toggle Amazon's monitoring of your reading habits. That's the information that amazon uses to compile your reading statistics that it displays in the kindle app, and to customize its marketing to you. By customized-marketing, I mean that Amazon will recommend similar books to those you've read in the past. Fantasy reader - Amazon will recommend more fantasy books. Have a favorite author - Amazon will show you their new book.

Tap the Disable button if you don't want this functionality.

9.3.6 Storage management

[3 vertical dots] > Settings > Device Options > Storage management

If you're Paperwhite is slowing down and being flaky, aside from needing a reboot, a full memory might be one reason why. In this area you can choose to upload content to the cloud and free up space.

- **Manual removal:** Remove content by type – Books, audible, periodicals, docs, samples.
- **Quick Archive:** Remove any items you haven't accessed recently – a month, 3 months, 6 months, a year.

9.3.7 Whispersync for Books

[3 vertical dots] > Settings > Device Options > Advanced Options > Whispersync for Books and notebooks

Tap *Whispersync for Books* to toggle Whispersync on and off.

Assuming that you're connected to both Wi-Fi and your Amazon account, Whispersync synchronizes your reading progress, notes, and highlights to your amazon account, and by default, to all your kindle devices. This means that you can pick up where you left off on one device and continue reading on another, and with your latest notes and highlights available. I always have whispersync enabled because its an important 'live' backup of my work.

If you're highlighting passages and writing notes, you should enable this.

I troubleshoot Whispersync in chapter 12.

9.3.8 Power Saver

[3 vertical dots] > Settings > Device Options > Power Saver

Tap Power Saver to toggle power saving functionality on and off.

- Toggle off to place your Paperwhite into standby mode so

that it turns on faster.

- Toggle on to place your Paperwhite into low power sleep mode.

After an extensive search of forums and blogs I can honestly say that what low power mode does is a mystery. The most likely thing it would do is to turn off background Wi-Fi.

Alternative (better documented) strategies for saving power include disabling page refresh, reducing the screen brightness (label (22) in Figure 5.3) and placing your kindle into airplane mode (label (17) in Figure 5.3).

9.3.9 Restart

[3 vertical dots] > Settings > Device Options > Restart

Tap Restart to, well, restart your Paperwhite. It's good practice to do this occasionally. It cleans up memory and can trigger it to check for updates. Sometimes restarting can also fix a flakey, slow or unresponsive device.

To restart your Paperwhite, tap Restart and in the box that appears tap the YES button. You can also restart your Paperwhite by pressing and holding the power button, and then tapping Restart in the box that appears.

9.3.10 Factory reset

[3 vertical dots] > Settings > Device Options > Factory reset

Tap Reset to restore your Paperwhite to its default factory settings. WARNING, all your personal data and content will be erased. This option could be useful if you were having problems with your Paperwhite and want to troubleshoot by restoring the default settings. A more likely scenario is that you want to sell your Paperwhite or give it away.

Assuming that just prior to the reset, you had Wi-Fi turned on so that your Paperwhite was synchronized to the cloud, when you setup your Paperwhite again, all your Books, notes and highlights (except content that you copied via cable) will be restored as a part of the setup process (chapter 4).

9.4 Screen and brightness

The Screen and brightness sub-menu lets control the appearance of your screen.

9.4.1 Show covers on lock screen

[3 vertical dots] > Settings > Screen and brightness > Show covers on lock screen

Tap Show covers on lock screen to toggle displaying either the cover of the book you're currently reading, or a screensaver.

I used to have this toggled on, but not anymore. I like to read widely and experiment with genres that I haven't read before. Recently, a book cover caught me off-guard. It was a bit too "racy" for my young daughter to see! I consider myself to be open minded, but when I looked closely, even I thought the cover was a bit challenging.

9.4.2 Appearance

[3 vertical dots] > Settings > Screen and brightness > Appearance

Toggle for dark mode (white text on black) or light mode (black text on white). You can also do this from the Quick Actions toolbar (label (19) in Figure 5.3).

9.4.3 Brightness and warmth

[3 vertical dots] > Settings > Screen and brightness > Brightness and warmth (see also Figure 5.3).

- **Screen brightness slider:** Tap and hold the slider, or tap the + or the – symbols for brightness values ranging from 0 to 24. Keep in mind that a very bright screen will discharge your battery faster.

- **Auto brightness:** Tap to allow your Paperwhite to automatically adjust the screen brightness to different situations (eg. reading under a tree versus reading in full sunlight). I show you how to calibrate this in 12.

- **Warmth slider:** Tap and hold the slider, or tap the + or the – symbols for warmth values ranging from 0 to 24.

- **Warmth Schedule:** Toggle the warmth levels according to a schedule. Automatically by sunrise and sunset, or manually between two times you choose.

9.4.4 Layout size

[3 vertical dots] > Settings > Screen and brightness > Layout size

Choose either standard size or large size layouts. I experimented with this and couldn't notice much difference. The large size seemed to make the text bolder.

9.4.5 Night mode brightness

[3 vertical dots] > Settings > Screen and brightness > Reduce brightness at night

Toggle to reduce the screen brightness as it gets darker. There is a light sensor at the top-center of the screen (see Figure 5.2).

9.5 Accessibility

[3 vertical dots] > Settings > Accessibility

Tap the Accessibility sub-menu to configure your Paperwhite's VoiceView Screen Reader text-to-speech tool, and change the layout size.

9.5.1 VoiceView Screen Reader

[3 vertical dots] > Settings > Accessibility > VoiceView Screen Reader

Tap VoiceView Screen Reader to have menu items and books read to you.

To enable VoiceView you need to first pair your Kindle to a Bluetooth audio device (headphones, speaker, etc). This can be tricky in the absence of being shown the exact process. I used my intuition and toggled the reader on, and tried to connect from the box that launched there. I WAS WRONG!!! I eventually got it to work, but I found the process frustrating, fickle and clunky. I was testing with late model Bose headphones, and two types of ear plugs.

This feature works better on later model kindle devices, most likely due to their faster processor.

Here's how to pair your blue tooth audio device...

1. Set your audio device into pairing mode.

2. Toggle VoiceView Screen Reader on and a message box will appear saying that your Bluetooth audio device has been recognized.

3. Tap two fingers on the screen for 1 second (index and middle finger worked best for me).

4. Wait and follow any further instructions.

5. After initial pairing, your device does not need to go through subsequent pairings.

Toggle the VoiceView Screen Reader button on to hear kindle text being read to you. When you toggle the button to be on, the tutorial, speech rate, volume and Bluetooth devices menus are enabled. The voice is a bit wooden, a problem that I imagine will be fixed as the current generation of AI voices matures.

Now when you touch a menu item, your kindle will read it to you. First it reads the name of the menu, and then it asks you to tap twice to select the item. I found that I needed to wait until the "tap twice" instruction finished before successfully double-tapping the menu item. It's a bit faster on the Paperwhite than on the paperwhite. When you're in a book VoiceView will also read it to you.

There are videos that demonstrate this on YouTube. Search for "voiceview screen reader kindle Paperwhite".

For example the @readerzplanet video at youtube.com/watch ?v=2cURLPUFLJU

Can tell that I'm not a fan of VoiceView? I did use words like "fickle, clunky and frustrating". You might find it more useful to explore text-to-speech in the kindle app on your smart device (enable text-to-speech in it's accessibility menu).

9.5.1.1 Tutorial

[3 vertical dots] > Settings > Accessibility > VoiceView Screen Reader > Tutorial

When VoiceView is toggled on additional menus appear

Tap Tutorial to access the four VoiceView tutorials ...

1. Basic gestures
2. Text entry with VoiceView
3. Advanced gestures
4. General information.

9.5.1.2 Volume

[3 vertical dots] > Settings > Accessibility > VoiceView Screen Reader > Volume

Tap Volume to adjust the volume on the VoiceView screen reader. Toggle to one of the eight volume levels. On my Bluetooth headphones, the volume control did not override this, and the up and down volume buttons toggled muting on and off instead.

9.5.1.3 Speech rate

[3 vertical dots] > Settings > Accessibility > VoiceView Screen Reader > Speech Rate

Tap Speech Rate to choose one of the ten speech rates on the VoiceView screen reader. There are two slower speeds, one normal speed, and seven faster speeds.

9.5.1.4 Bluetooth Devices

[3 vertical dots] > Settings > Accessibility > VoiceView Screen Reader > Bluetooth Devices

Tap Bluetooth Devices to change to another Bluetooth audio device that's already been paired.

9.5.2 Appearance

[3 vertical dots] > Settings > Accessibility > Appearance

Tap Invert Black and White to toggle between white screen with black writing and black screen with white writing. You can also do this from the Quick Actions toolbar (label (19) in Figure 5.3).

9.6 Home, Library and Notebooks

[3 vertical dots] > Settings > Home, Library and Notebooks

Tap Home and Library to choose how your books and collections are displayed. You can toggle to...

- group book series together.

- choose how you want your collections to display (3 options).

9.6.1 Library tab

[3 vertical dots] > Settings > Home, Library and Notebooks > Library tab

In the Library tab you can. . .

- **Collections:** Toggle to control the way collections are displayed (Always show, Show favorites, Show only in collections view)

- **Audible:** Toggle to Always show or Show only when filtered

9.6.2 Book and notebook options

[3 vertical dots] > Settings > Home, Library and Notebooks > Book and notebook options

From the Reading Options sub-menu, you can toggle Page Refresh and Vocabulary Builder on and off.

9.6.2.1 Page refresh

[3 vertical dots] > Settings > Home, Library and Notebooks > Book and notebook options > Page Refresh

Tap to toggle Page Refresh on and off. Page refresh overcomes the ghosting problem that can occur on e-ink devices. Ghosting is where bits of the previous page you were reading still display on your screen, and make it look smudgy.

Turning Page Refresh off can also be a component of a power saving strategy. Personally, my priority is to have a screen that's clear, bright and easy on my eyes. I always have access to recharging facilities (even when bush camping), so I've never bothered about this power saving option. Unless you are in a situation where it's essential to conserve power, my advice is to toggle Page refresh on.

9.6.2.2 Vocabulary Builder

[3 vertical dots] > Settings > Home, Library and Notebooks > Book and notebook options > Vocabulary Builder

Tap Vocabulary Builder to toggle it on and off. Vocabulary builder records words that you've looked up in your Paperwhite's dictionary. The idea is that every now and then you'll access the tool and revise the words you've looked up. I explain how to use it in chapter 6.

I don't use this functionally often, but I do like it. I imagine that it would be very useful to students and language learners.

Vocabulary Builder is limited to 2000 words, so older words get removed once the limit is reached. The word list is not updated if you disable this feature.

9.7 Help

[3 vertical dots] > Settings > Help

There's four types of help in this menu...

- **Get Started:** This is a quick guide to setting up your Paperwhite and interacting with the hardware.

- **Kindle User Guide:** Use as an extra resource to this book. This is a comprehensive guide to your Paperwhite. Much like the eBook version of this book, it has headings with links. If I've written about something and you don't understand it, this manual is a great place to read someone else's take on your problem. It is surprising that there is no ability to search for words that might describe your problem.

- **Tutorials:** Contains an interactive navigation tutorial. Worth taking. You'll also find the labelled screen captures at bit.ly/akpug a useful resource to keep by your side while you're reading this book.

- **Contact us:** various problem focused menus with a phone button that leads to contact phone numbers in your local region. You may need information from the Device Info area.

9.8 Legal and compliance

[3 vertical dots] > Settings > Legal and compliance

Yep. If you're having problems getting to sleep, there's legal and compliance reading for you! They outline, of course, the rights that amazon says you have. But that's the same with all license agreements we regularly sign up to isn't it!

Chapter 10

Managing your Kindle from a web browser

You can manage your Scribe, access your reading history, and review your notes and highlights when you're logged into your local Amazon store in a web browser. Some tasks require the use of the Amazon website in a browser, while tasks like editing notes and highlights are better suited to the browser-based kindle cloud reader.

This chapter is about the amazon web site on a desktop web browser. If you're using a browser on a smart device, depending on the device, some of the functionality I talk about may be missing.

10.1 Manage your Paperwhite from your local amazon account

amazon.com/myk [amazon.co/uk/myk, amazon.ca/myk, amazon.com.au/myk]

In the manage content and devices area of your local store, amongst many other things, you can install additional language dictionaries, and deregister your Paperwhite so you can sell it. You can also view the content on your Paperwhite, including your entire Kindle Unlimited and Prime reading borrowing history.

Since the first edition of this book, amazon has created a new way to access this - amazon.com/myk. You need to be in your local store to get to access to the full range of management options (ie. amazon.co.uk/myk, amazon.com.au/myk, amazo n.ca/myk, amazon.com.au/myk, etc).

10.1.1 Install a different language dictionary

Amazon in a web browser > Accounts & Lists drop-down menu > Manage content and devices > Content > Dictionaries and User Guides

OR

amazon.com/myk > Content

From the Manage content and devices area of your amazon account, click Content, then Dictionaries and User Guides to view the 50+ available dictionaries. Select a dictionary and then choose to Deliver to Device.

10.1.2 Manage kindle devices

Amazon in a web browser > Accounts and Lists > Manage content and devices > Devices

OR

amazon.com/myk > Devices > Amazon devices

You can manage your kindle devices (Paperwhite, Paperwhite, Fire, etc) and apps (oasis, desktop, smartphone, etc.) from the devices area of your local amazon account. In a browser, login to your account, and follow the path above. Click the large Kindle button and you'll see your Paperwhite listed.

Choose your Paperwhite and you can view its model and serial number, change its label (eg. "Paperwhite" to "Ian's Paperwhite"), deregister it (if you sell it, or it was stolen), set it as the default device that all content will be automatically delivered to, and view all the content on it (books, documents and dictionaries).

10.1.3 Manage Kindle orders

Amazon in a web browser > Accounts and lists > Manage content and devices > Content

OR

amazon.com/myk > Content

In a browser, login to your amazon account, and follow one of the links above.

Here you'll find a list of all the books you've ever bought, and active Kindle Unlimited and Prime reading loans. It's where you manage your books – return a loan, deliver a book to or remove a book from any of your kindle devices, add or remove a book from a collection, clear the furthest page read, mark a book as read, and view the details of a book order.

If you're a Kindle Unlimited subscriber or Prime member, by default, you're shown your current borrowings. You can also view books that you've returned. To do this, choose to view Kindle Unlimited [or Prime Reading] from the drop-down menu just above your books, and then Returned from the drop-down menu to its right.

10.1.4 Cancel or pause Kindle Unlimited

Amazon in a web browser > Accounts & Lists drop-down menu > Subscriptions > Your membership and Subscriptions > Kindle Unlimited > Kindle Unlimited Settings

OR

amazon.com/myk > Preferences> Kindle Unlimited Settings > Manage your subscription

In your browser, login to your local amazon account, and follow one of the paths above.

When you cancel your subscription, books that you've borrowed will be removed from your Paperwhite the next time it connects to Wi-Fi. So, if there's a Kindle Unlimited book that you really want to finish, be sure to place your Paperwhite into airplane mode (label (18) in Figure 5 3) before you cancel your subscription.

Amazon realizes that people hop in and out of subscription services, so if you resubscribe in the future your Kindle Unlimited borrows, and notes and highlights will be reinstated.

10.2 The Kindle cloud reader

read.amazon.com/

The kindle cloud reader lets you read any kindle book in your library in a web browser (except documents that you've uploaded via cable). To use it, sign into your Amazon account and go to read.amazon.com.

The kindle cloud reader has much better note taking and highlighting functionality than your Paperwhite. Rather than using my stumpy fingers to take detailed notes on my Paperwhite, I find it easier to use a keyboard and mouse to correct, expand on and color-code my notes.

The Kindle Cloud Reader has a sidebar and a top bar. These change depending on whether you're in your library or reading a book.

10.2.1 Kindle cloud reader - library mode

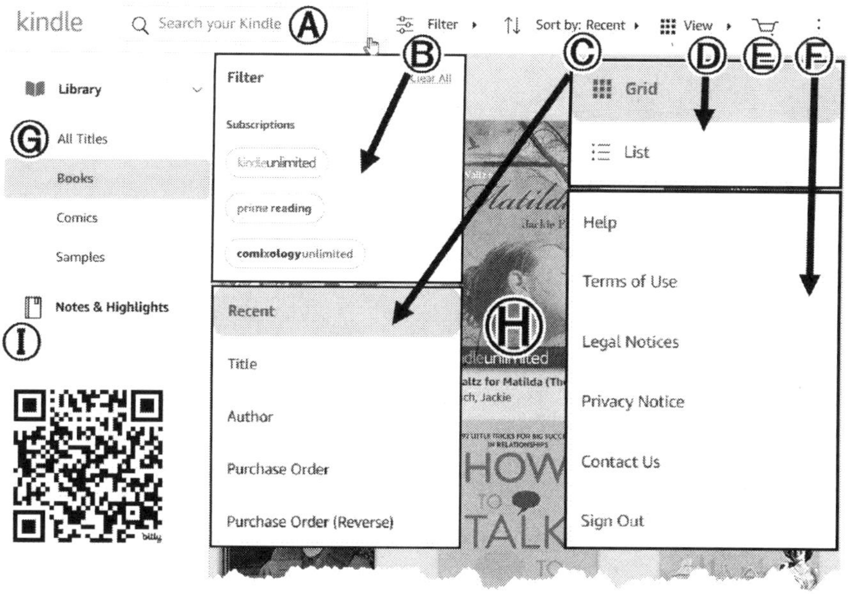

Figure 10.1: The Kindle cloud reader - library mode (download the figures from bit.ly/akpug or the QR code).

While you're in Library mode, the top menu bar contains options for filtering your books by subscription type, sorting, grid view and list view, and kindle store. The sidebar has options to display either books or notes. The main area displays all the kindle books in your library that you've purchased, or currently have on loan from Prime Reading, Kindle Unlimited or ComiXology Unlimited.

The menus and options available in the Kindle Cloud Reader are shown in Figure 10.1 and include. . .

Label (A) Search box: Type the name of any book in your library into the search box to find it.

Label (B) Filter: Filter books according to your subscription.

Label (C) Sort: Sort your books in various ways – recent, title, author, or order of purchase.

Label (D) Display: Toggle to display as either a grid or a list.

Label (E) Kindle store: Click the shopping trolley to be taken to the Kindle store. When you buy a book from here, its delivered to your cloud reader. You can also choose to deliver them to any of your kindle devices and apps from the drop-down menu on the store's thankyou page.

Label (F) Help, notices, and sign out: This menu provides access to resources and support for the Kindle Cloud Reader, including a user guide, FAQs, and contact information for customer service.

Label (G) Side bar: In the side bar you can elect to show all titles or filter kindle books, ComiXology books and samples.

Label (H) The main area: This is where all the books in your library are displayed.

Label (I) Notes and highlights: Here's where you view the notes (except handwritten notes) and highlights that you made in books that you've read, including for kindle unlimited and prime reading books that you've returned. Clicking on a book launches the box shown in Figure 10.2.

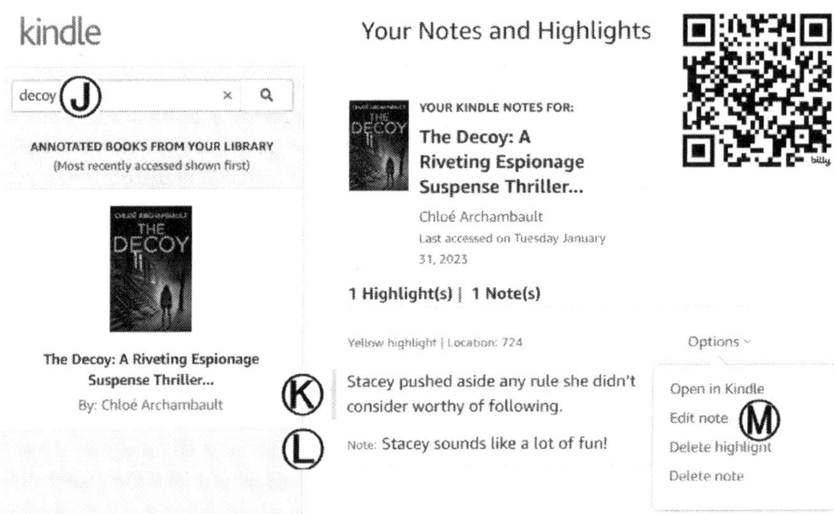

Figure 10.2: Notes and highlights in the Kindle cloud reader (download the figures from bit.ly/akpug or the QR code).

When you're in Notes and Highlights mode, a list of all the titles you've added notes and highlights to, appears in the sidebar. When you tap on a thumbnail, your notes and highlights for the book appear in the body of the screen.

I have to say that only now, as I'm writing this, do I realize what amazing functionality this is. It hadn't occurred to me that my notes and highlights would build up over time. I can't resist giving you a few examples. . .

> "Leave pettiness to the petty". Jay Christoph. Nevernight.

> "Men could be wonderful fathers. Sarah just hadn't picked the right one." Melinda Leigh, She can kill.

> "I'd like to say that Natalie isn't pretty. But that would be a lie . . . It's only when she opens her mouth that she turns ugly." Suzanne Redfern, In an Instant

> "My driving instructor, a bald man with severe halitosis and nerves of steel labelled the impediment 'Pedal dyslexia' ". Suzanne Redfern, In an Instant.

I am having such a blast catching up on the essence of my years of Kindle reading. There are so many one-liners that I had forgotten.

Best I get back to writing his book. Let's talk about the labels in Figure 10.2.

Label (J) Search: You can type the name of a book to search for in your library. Even Kindle Unlimited and Prime reading books that you've returned.

Label (K) Highlight words and phrases in a book: *Stacey pushed aside any rule she didn't consider worthy of following* is a sentence that I highlighted.

Label (L) Comment on a highlighted word or phrase: *Stacey sounds like a lot of fun* is a comment I made about the text that I highlighted.

Label (M) Note options: From the options menu you can choose to open the book in kindle, edit the note, delete the highlight or delete the note.

10.2.2 Kindle cloud reader - book mode

The Kindle Cloud Reader offers many of the same features as the Kindle app and Paperwhite, including the ability to adjust font size, background color, note taking, book marking and dictionaries. Click on a book's thumbnail to open it. The book mode menu is labelled in Figure 10.3.

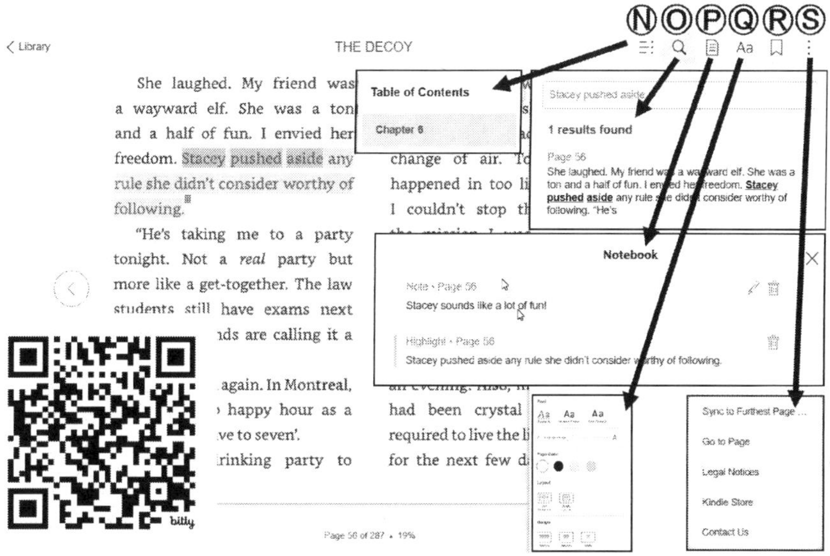

Figure 10.3: The Kindle cloud reader menus (download the figures from bit.ly/akpug or the QR code).

Label (N) Table of contents.

Label (O) Search: Search for a word or phrase within a book.

Label (P) Notebook - Highlight and annotate: Highlight and annotate text and edit your annotations. You can also choose to color code your note yellow, blue, pink or orange.

Label (Q) Font: Change the font, its size, page color and layout of the page.

Label (R) Bookmark: Add or remove a bookmark.

Label (S) Additional menu: Sync to the cloud, go to a page, legal notices, kindle store and contact amazon.

Chapter 11

Subscriptions and Amazon store book categories

Amazon in a web browser > Kindle Store > Kindle books

You should avoid going on a book buying spree until you understand Amazon's book subscriptions, and book categories. Understanding these has expanded my reading horizons – it's helped me read more widely, more often, and more economically than I ever did. Who knows what might happen for you?

Amazon's subscriptions include Kindle Unlimited and Prime. Major book categories include Kindle Short Reads, Kindle Vella (retired in February 2025), Kindle First Reads, Kindle Singles, Amazon Charts and eBooks that come bundled with audible. You subscribe to these from a browser. Unfortunately, not all subscriptions and categories are available in every country.

11.1 Subscriptions

Given the choice between a Kindle Unlimited and Prime subscription, you may find that Prime is a better place to start. It's a bit cheaper than Kindle Unlimited, and has wide ranging benefits. As well as Prime viewing and Amazon music, and free deliveries on some goods, for readers it includes a subset of Kindle Unlimited, Graphic Novels and Audio. The kindle store on your Paperwhite and other devices automatically recognizes your subscriptions, and for 'eligible' books, the buy button changes to read now for $0.00.

11.1.1 Kindle unlimited

Kindle Unlimited is a subscription library service offered by Amazon that gives you access to eBooks, and audiobooks. By eligible, I mean a book where its publisher has enrolled it in Kindle Unlimited (they're paid ~$4 per 1000 pages read). To enroll, they must agree to making their book exclusive on amazon. There's around four million titles that are labeled as Kindle Unlimited in the kindle store. Eligible books are clearly labelled with Kindle Unlimited above their thumbnail.

When a kindle unlimited subscriber is logged in to their account, for eligible books, the text in the Buy button automatically changes to Read Now for $0, or to $0.00 Kindle Unlimited. Undoubtedly there'll be different versions of the button over time.

When you find an eligible book you want to read, tap on the book thumbnail to open its sales page and then tap the Read now button to download the book and start reading. You could also choose to buy the book if its one that you want to keep beyond your Kindle Unlimited membership.

The kindle unlimited sales page says that you can borrow 10 books at a time for a flat monthly fee - currently ~$10 US / month. Maybe its because me and my partner read lots, but I'm allowed to borrow 20 books. Once you reach your borrowing limit, you'll be prompted to return a book so you can borrow another. There is no such thing as a kindle yearly subscription. The monthly fee is debited automatically using the payment method nominated on your account.

From the Amazon store in your Paperwhite (label (3) in Figure 5 1), browser, or app, you can filter to show you only Kindle Unlimited eligible books.

You subscribe to Kindle Unlimited by logging into your Amazon account and then clicking on Kindle Unlimited in the side menu, or type Kindle Unlimited into the Amazon store's search bar and follow the prompts.

11.1.2 Prime Reading

Prime members have access to a subset of the Kindle Unlimited library, and one free First Reads book a month. There are 3000 or so regularly rotated titles to choose from, some of which come bundled with audio. Prime is a good way to get a taste for the Kindle Unlimited experience, especially seeing as you may already be a member for its delivery, movie, music and other benefits.

If you're a Prime member, in all the Kindle apps and Kindle stores there's a Prime Reading filter that you can tap to see only Prime results.

Amazon Prime Reading replaced the Kindle Owners Lending Library (KOLL) in 2020.

11.1.3 Kindle newsstand

Kindle Newstand was closed in mid-2023.

11.2 Featured categories

The featured categories in the kindle department in the browser version of the amazon store contain some novel ways of searching for your next read.

11.2.1 Narrated books

Amazon in a web browser > Kindle Store > eBooks with Audible

In your browser, login to your local amazon account, and from the dropdown menu attached to the search bar click Kindle Store and press enter, and then eBooks with Audible from the top menu. There's also a books with audible narration category in the kindle app.

Audible is not enabled on the Paperwhite in my marketplace, but it is linked within the Kindle app on my smart phone. Kindle books with narration have a headphone symbol next to the title and there's an "add audible narration" upgrade tick on the sales page. There are around 2600 titles (March 2023) where the Audible version is included for free to Kindle Unlimited subscribers. A small subset of these is available to Prime members.

In the kindle app on your smart device, tap the top of the screen for the standard toolbar. An Audible Narration – tap download link appears at the bottom of the screen. When you play the audio, the app highlights each word as it is being read. What a great tool for language learners!

You can use Amazon Matchmaker to interrogate all the books in your library and match them with an audible narration if one is available. Audio books discovered by Matchmaker are often a fraction of the stand alone audible price. You need to be logged in to your amazon account and on a browser. Google "Amazon matchmaker" to find it.

11.2.2 Kindle short reads

Amazon in a web browser > Kindle Store > Kindle Short Reads

In your browser, login to your local amazon account, and from the dropdown menu next to the search bar click Kindle Store and press enter. Then click on Kindle Short Reads in the side menu. Otherwise, type short reads into the search bar and it should appear at the top of the search results.

Short reads are kindle books categorized on amazon's website by page length and estimated read time – 15, 30, 45, 60, 90, and 120 minutes.

Sometimes this category can be hard to find. If all else fails, google Kindle Short Reads.

11.2.3 Amazon first reads

Amazon.com/firstreads

In your browser, add '/firstreads' to the URL of your local amazon store. Otherwise, type first reads into the amazon search bar and it should appear at the top of the search results.

First reads allows you to download, read and keep one new pre-release book every month. It provides access to a selection of new books that have not yet been released to the general public. Each book, written by well-known authors, is $2 from the amazon storefront, or free for Prime members from either the storefront or the app.

You have to be a First Reads member to gain access though. Joining is free, but by joining, you're giving amazon permission to send you a monthly email about the program. There's two ways to participate. . .

- **Prime members:** Prime members can sign up to get one first-read per month for free.

- **Non-members:** Join Amazon first reads from the first reads area of the kindle store. You need to be logged in to your local Amazon store to do this. Once you're joined, you can buy one first release book per month for $2.

11.2.4 Kindle singles

Amazon in a web browser > Kindle Store > Kindle Books > Kindle Singles

In your browser, login to your local amazon account, and from the dropdown menu attached to the search bar click Kindle Store and press enter, then Kindle Books, and then, in the side bar Kindle Singles.

Kindle Singles are novellas and essays that are usually shorter than traditional books, but longer than magazine articles. They are often written by well-known authors, journalists, or experts in their field.

The category is obvious in the browser version of the amazon store, but can be hard to find on your Paperwhite. It's a link at the very bottom of the Kindle Store on your Paperwhite.

11.2.5 Kindle Vella (serials)

Amazon in a web browser > Kindle Store > Kindle Books > Kindle Vella

In your browser, login to your local amazon account, and from the dropdown menu attached to the search bar click Kindle Store and press enter, then Kindle Books, and then, in the top bar Kindle Vella. If all else fails, google Kindle Vella.

Kindle Vella is serialized fiction. You can follow stories in bite-sized installments, with new chapters being released on a regular schedule.

The first few episodes of every story are free. You unlock later episodes with "Tokens" that can only be used in Vella. See more at www.amazon.com/kindle-vella.

Unfortunately, Kindle Vella is only available in the US for people with a US account.

* Retired in February 2025. Try wattpad.com as an alternative serialized fiction site.

Could you spare a moment to review my book?

Thank you so much for reading my book! If you've found it helpful, I'd be incredibly grateful if you could leave a short review. Reviews are a huge help to indie authors. They help other readers discover our books and keep us writing. Just a few words about what you liked would mean the world. Thanks again for your support!

Review on Amazon US		Review on Amazon UK	
 https://amzn.to/4hFb80w **Click the Flag or Link**	**Scan the QR code**	 https://amzn.to/3EA7u9K **Click the Flag or Link**	**Scan the QR code**
Review on Amazon Canada		**Review on Amazon Australia**	
 https://amzn.to/3Equs3q **Click the Flag or Link**	**Scan the QR code**	 https://amzn.to/4hmKEkv **Click the Flag or Link**	**Scan the QR code**

Don't forget the resources from the QR codes in the figures, and at ianallanauthor.com/paperwhite

Chapter 12

Troubleshooting

If you're here for a reason other than curiosity, its likely that you're anxious, frustrated and possibly a little angry. Take it from someone who, in a former life, was responsible for supporting the foreign exchange area of a major bank, maintaining calm in your crisis is your friend. Be methodical and document your steps, and you're far more likely to solve your problem. So, take a deep breath and read on. . .

12.1 Setup problems

12.1.1 I accidently setup in a foreign language

If you accidently chose a foreign language during setup, and you can't read the menus to understand where reset is, here's the menu path...

More (bottom-right of screen - (11) in Figure 5.1) > [3 vertical dots] > Settings (4th item down) > Device Options (3rd item down) > Factory reset (last item) > then tap YES (bottom right button in the box)

12.1.2 PIN problems

You cannot reset your PIN if you've forgotten it. You'll need to reset your Paperwhite.

- Enter 111222777 into the passcode area to erase your Paperwhite and restore the factory settings.
- Once reset, follow the prompts to get connected again (chapter 4), and your kindle will re-populate from the cloud, but without a PIN. This may take a while if you have lots of books to download.

12.2 Content problems

12.2.1 Strange symbols in books

There are two likely reasons for strange symbols in your book. . .

- **Wrong language setting:** The language of your Paperwhite may not be set to the language the book was written in. Change your Paperwhite's language in the Languages and Dictionaries area of the Settings menu (chapter 9).
- **Author error:** The author may have written the book in an incompatible word processor. For example, kindle does not recognize some of the smart formatting (eg smart quotes) in MS word, and displays them as strange symbols.

12.2.2 What happens to my books if get a new Paperwhite or I lose my old one?

Browser: Amazon > Your account > Manage your content and devices > Devices > [select your paperwhite] > Settings > Set as default

OR

Paperwhite: Swipe down for the Quick Actions toolbar > Sync (label (20) in Figure 5.3)

All your kindle purchases are attached to your amazon account in the cloud, and so, aside from content that you've transferred to your Paperwhite via cable, none of your books will be lost. When you connect your Paperwhite to your amazon account during setup, all your content is downloaded and synchronized with all your kindle devices. That includes notes, bookmarks, and even the page you last read. Initially only thumbnails are downloaded, so you'll need to tap on a book's thumbnail to download the entire book.

If your books don't automatically appear, try the following things. . .

- **In a browser, set your Paperwhite as the default device:** From your local amazon web page, click Your account, then Manage your content and devices, then in the Amazon Devices tab, click on your Paperwhite. In the settings area of the screen that launches, click the Set as default device button.

- **In a browser, manually deliver one or all of your books to your Paperwhite:** From your local amazon web page, click Your account, then Manage your content and devices, then from the Content tab, select one or all of your books. Then click the Deliver to Device button and choose your

Paperwhite from the list (chapter 8).

- **On your Paperwhite:** Swipe down from the top of any screen to access the quick actions toolbar. Tap the synchronize icon (label (20) in Figure 5.3) to manually synchronize your Paperwhite with all your kindle apps and devices.

Be patient. Sometimes Amazon can take a little while to recognize a new device and sync it. By a little while, I mean say, 10-15 minutes. Sometimes it can help to reboot your new Paperwhite to force it to update and check the amazon cloud for new content (chapter 4).

Exceptions are books that you have transferred to your kindle via USB cable. That's because cable-transfers are not backed up in the cloud - another reason to use sendtokindle.

12.2.3 Why won't my book transfer via cable?

There is a disconnect between the many "use a USB cable to get a book on your Paperwhite" style blog posts, and reality. It's not as simple as some bloggers make it out to be. Or maybe they just have a well-oiled workflow that works for them? There are two reasons why cable transfer often does not work, and sometimes it can be hard to tell which of the two it is.

- **You might have a corrupt eBook file:** E-book formatting is a temperamental and fickle art. Even people who create eBooks all the time dread the kindle compatibility testing stage. That's because every book creation software creates eBooks slightly differently. Something as simple as one incompatible character can cause an eBook to fail. If your Paperwhite doesn't recognize the file that you're transferring, try importing it into the free Calibre software (google

Calibre), then export the file as an ePub and transfer that. Take it from someone who's done lots of data-cleaning work over the years, sometimes using different software is the easiest way to overcome data compatibility problems.

- **File location:** Different bloggers suggest copying your book to different locations on your Paperwhite. The documents folder, and a folder inside the documents folder with the same name as your file are two common recommendations. For me, some of those locations worked for some files, and not for others.

I recommend that you do not use the cable transfer technique. Use the send to kindle drag-and-drop or email techniques instead (chapter 8).

12.2.4 Why won't the book I bought show up on my Paperwhite?

Amazon in a web browser > Accounts & Lists drop-down menu > Manage your content and devices > Books > Deliver or Remove from device

OR

Amazon.com/myk > Manage your content and devices > Books > Deliver or Remove from device

Sometimes Amazon delivers your book to the wrong device, especially if you have lots of kindle apps and devices. So, you may need to manually deliver it to your Paperwhite. From your local amazon web page, click Your account, then Manage your content and devices, then from the Content tab, select one or all of your books and then click the Deliver to Device button and choose your Paperwhite from the list. You may also need to make sure that your Paperwhite is nominated to be your default-device (see Manage kindle devices in chapter 10).

12.2.5 Why won't the file I'm emailing appear on my Paperwhite?

If you've emailed a file and it's not appearing on your Paperwhite, check for the following problems. . .

- **Unauthorized email address:** Your kindle will only accept emails from an authorized address. So, if you haven't already, you'll need to go to your amazon account and add the email address you're sending from (chapter 8).

- **Wrong kindle email:** Are you sending to the correct kindle address? Kindle email addresses often contain the number zero and the letter O. You may be typing O's instead of zeros or visa versa? The letter O is fat and the number 0 is skinny.

- **Wi-Fi:** Do you have Wi-Fi signal?

- **Wrong file type:** Are you sending a file type that your Paperwhite can read? Valid file type are .azw, .epub, .mobi, .PDF, .doc, .docx, .rtf, and .html document files. Also .png, .jpg, .gif and .bmp image files.

- **Corrupt file:** The file you're sending might be corrupt. Try importing it into the free Calibre software (google Calibre), then exporting it as an ePub file. Now email the ePub that you just exported.

- **Reboot your Paperwhite:** Sometimes rebooting your Paperwhite can help. Rebooting forces your Paperwhite to update and check the cloud for new content (chapter 4).

12.2.6 Why won't my book synchronize to all my devices?

Sometimes Whispersync needs to be given a manual nudge to get all your devices synchronized. For example, you may have borrowed a book on one device, returned it on a different device while it was open on a different device again, and then borrowed it again (yes, that can happen). From any place on your Paperwhite, swipe down for the Quick Actions toolbar and then tap Sync (label (20) in Figure 5.3).

12.3 Charging problems

If there's no green light or amber light near the charge point, it means that your Paperwhite is not charging. Here's some things to check. . .

- **Are all the connections clean:** Dust and grit can get into the charge points on both your charger and your Paperwhite. Dust the socket connection on your Paperwhite with an artist's brush or do a hard (spit-free) mouth-blow into it. Clean the pin connection on the cable with a soft cloth.
- **Is the cable inserted properly:** Make sure it's not loose.
- **Faulty cable:** Try a different one.
- **Faulty charger:** Try a different one.
- **Charger type:** Even though USB-C is an adaptive power protocol, it's wise to avoid no-name chargers, and chargers

> 9w (especially ones labelled fast charger or similar). If you do use a higher power USB-C charger, be sure to supervise it closely the first few times you use it.

- **Computer USB port:** Make sure that you're not using a low power USB port. Low power ports are common in peripherals such as keyboards. Swap to a different charging port.

12.4 Hardware problems

12.4.1 Why does my battery drain so quickly?

Amazon's battery life claims are based on a half hour of daily reading, and your Paperwhite being in airplane mode, and with the light setting at level 13. I read for 2+ hours most days, airplane mode is never enabled, and my screen brightness is set at 20. I charge my Paperwhite most weeks. My partner reads for 2+ hours most days, in dark mode at night, airplane mode is never enabled, her screen brightness is set at 18, and she often uses a large font (and so has more page turns). She charges her gen 11 paperwhite once a month.

If your battery life cannot be explained by the way you use your Paperwhite, then check these things. . .

- **Cover faulty:** If you're using a cover, is it somehow interfering with sleep mode and causing your Paperwhite to turn on and off all the time?

- **Cover for wrong Paperwhite model:** Are you using a cover that's designed for your Paperwhite model? When you're buying a cover, be sure to check that its specifications match your kindle model. A mismatch might cause sleep mode to malfunction and drain your battery.

- **Storing on the power button:** Are you storing your Paperwhite on its side (eg. in your bag) so that the power button is flicking on and off without realizing it?
- **Brightness too high:** Lower your screen brightness, but only if its comfortable for your eyes (chapter 5).
- **Wi-Fi issue:** You could have a "confused" Wi-Fi connection that's causing excessive communication between your Paperwhite and your router. Reset your Wi-Fi by turning the router off. Remove the plug from the power outlet for a minute or so, so that it completely drains of power, and then reconnect it. Then check that your Paperwhite is properly connected to your router (see the Why can't I get my Wi-Fi to work heading below).
- **Confused operating system:** All devices perform better if they're restarted occasionally. So, restart your Paperwhite (chapter 4).
- **Overcharged battery:** You might have overcharged your Paperwhite. This is less likely than it used to be, but still not impossible.

As with any battery device, consistently overcharging your Paperwhite (ie. leaving it charging long after it reaches 100%), will shorten its battery life. Some commentators suggest that its good practice to run the battery completely down once every month or so.

12.4.2 Why don't my screen taps work?

Sometimes you can tap multiple times thinking that your tap has not registered, and then your Scribe flicks through multiple pages. Most likely that's because your Scribe is busy with the cloud while you're tapping. I had this problem often on my 10th paperwhite. My December 2024 12th generation paperwhite is lightening fast, so I no longer have this problem.

Other times, where processor speed is not the issue, it simply won't respond to your taps and swipes. Here's some things to think about. . .

- **It might be because your fingertip is dry or calloused:** Paperwhites use capacitive touch screen technology. That means that your finger completes an electrical circuit when you touch the screen. So, a dry or calloused finger will act as an insulator rather than as a conductor. Here's some simple things to try.

    - **Dampen your fingertip:** You could use a finger-moistener-sponge like bank tellers and postal workers use. Search online for "sponge cup" or "finger moistener". Or, you could just lick your finger.

    - **Use a different part of your finger:** Try the side of your finger instead of the tip.

    - **Use a different finger:** Some fingers might be better conductors than others.

    - **Use a lighter touch:** You might be pressing too hard.

- **Is your screen dirty?** Clean a dirty screen with a soft tissue or moist cloth.

- **Is your battery low?** A low battery can cause problems on any device. Or, Maybe your outlet, cable or charger is faulty and so you think your Paperwhite is charged when it is not. Does the orange charging light appear when its plugged in? One piece of hardware at a time (power outlet, cable, charger), try swapping them out.
- **Is the charge port dirty?** Clean it. Often a strong blow from your mouth is enough to dislodge dust and dirt.
- **Is your Paperwhite too hot?** Most devices go into protection mode if they're left in the hot sun. Rest your Paperwhite somewhere where it can cool down and try again later.
- **Is your screen bright enough for you to see the text?** Adjust it to be brighter (label (22) in Figure 5.3).
- **If all those suggestions fail:** First, you should try restarting or resetting your Paperwhite (chapter 4)
- **Get a new Paperwhite:** If none of these suggestions work, then you may need to replace your Paperwhite. . .
    - **Within warranty:** Contact amazon support. Carefully document the troubleshooting steps you've taken. Hopefully they will replace it.
    - **Damaged outside warranty conditions**: If something happened like your Paperwhite was dropped or sat on, or the USB port is physically damaged, you may need to buy a new one.

12.4.3 Why can't I get my Wi-Fi to work?

[3 vertical dots] > Settings > Wireless > Wi-Fi networks

I know. It would be easy to have a meltdown when you're troubleshooting a Wi-Fi connection. But it's important to be calm and methodical. Look for the most obvious problems first. Here's some things to check. . .

- **Are you in Wi-Fi range?** Make sure your Paperwhite is within range of a Wi-Fi network, that there's strong signal, and that the network is functioning properly (it has power and is connected to the internet). You can check this by using the web on another device that you know is connected to the same band of the same network you're trying to connect to (ie. most Wi-Fi has 2.4Ghz and 5Ghz bands). Place your Paperwhite next to the other device while you test both. If the other device is working and your Paperwhite is not, then go to the next step.

- **Are you connected to the correct Wi-Fi?** Make sure your Paperwhite is connected to the correct Wi-Fi network – the one you just tested. Select Settings from within the More menu (label (11) in Figure 5.1), tap Wireless and then tap Wi-Fi Networks. Tap on the correct network if you need to.

- **Wi-Fi won't accept your password:** If the network won't accept your password . . .

    - **Are you trying to connect to the correct network?**

    - **Are you typing the password correctly?** Passwords are often case sensitive. And, have you accidently confused the capital letter O and the number zero? The number 0 is skinny and the letter O is fat. Take a photo of the password with your smartphone. That way you'll be able to zoom in to see the small text.

- **You can't find the Wi-Fi password:** Network providers often include a fridge magnet with your Wi-Fi name and password printed on it. Look on your refrigerator (front and sides). Otherwise, look for a sticker on the side of the Wi-Fi router. Take a photo of it with your smartphone. That way you'll be able to zoom in to see the small text.

- **Restart your Wi-Fi:** Restart your Wi-Fi router by unplugging it from the power source for a minute or so, and then plugging it back in. Be sure to remove the plug from the power outlet completely. For some routers, just using the on/off switch is not enough. Removing the plug from the power outlet and waiting for the power to drain from the device can force it to do a full reset and check for firmware updates.

- **Reset your Wi-Fi:** Unfold a paperclip or something similar and poke it into the tiny hole on your router labelled "reset". Google "how to reset [your router make and model]" for exact instructions.

- **Your router frequency could be incompatible:** Check what frequency your Wi-Fi router is operating in. Paperwhites operate in both the 2.4 Ghz and 5 Ghz frequencies. Earlier model Kindles only operate in the 2.4 Ghz frequency.

- **You might be too far away from your router:** The 5ghz frequency is fast and works well over short distances. In my home, the router is downstairs and we have problems receiving the 5ghz frequency upstairs on the opposite end of the house. We have no problems with the 2.4ghz frequency - its slower, but it copes better with thick walls, and longer distances.

- **Try a smartphone hotspot:** Create a Wi-Fi hotspot with your smartphone and test if you can connect to that.
    - **iPhone:** settings > personal hotspot
    - **Android:** settings > network and internet > hotspot & tethering > Wi-Fi hotspot
- **You may be trying to connect to the wrong frequency:** Newer Wi-Fi routers operate in the 2.4Ghz and 5/6Ghz frequencies simultaneously. For older Kindles, check that you're trying to connect to the 2.4Ghz frequency.
- **Your router is old:** Some older Wi-Fi routers do not support the 2.4Ghz frequency. Google the make and model of your router to find this out. Replace your router if you need to.
- **Your router is in a hidden network:** Some networks operate in hidden mode. Add these wireless networks manually if a rescan does not detect them. To find your router's IP address, google how to find my router IP address on [smartphone | windows | iOS | android].

If all of these suggestions have failed and your Wi-Fi will not connect to your Paperwhite, or any other device, contact your internet service provider for assistance. Otherwise, you may need to try a forum such as reddit.com/r/kindle/. It will help people to help you if you have documented the steps you've taken, and EXACTLY how you went about them (including in what order).

12.4.4 Auto brightness doesn't work

If the auto brightness doesn't work (label (23) in Figure 5.3), you may need to calibrate it. The Paperwhite has a light sensor at its top.

To calibrate your Paperwhite, place it in full sunlight (with the sensors unobstructed). With auto-brightness toggled on, slide the brightness slider to zero. You could also use a lamp or even the torch from your smartphone as a light source.

12.4.5 Contact Amazon support

If all else fails, you may need to contact support. Look for Contact us in the Settings menu. First check that your firmware version is up to date. You find that information in the Device Info area of the Settings menu.

Chapter 13

Frequently Asked Questions

Here's some answers to common questions about the Paperwhite. I found them by researching forums, amazon product pages, and amazon help pages.

13.1 Content

13.1.1 Does a Paperwhite need Wi-Fi to work?

Your paperwhite does not need Wi-Fi if the book you're reading is already on your device. However, you need to keep in mind that the kindle is a device that's designed to give you access to Amazon's Kindle Book Store. It's there to sell amazon books!. You cannot download books from the kindle store to your Paperwhite if you don't have Wi-Fi. That's because books from the kindle store are protected by Digital Rights Management (DRM).

Without Wi-Fi you can connect your computer to your Paperwhite via cable, and then transfer eBooks and documents that you've sourced from non-amazon sites. I found that process to be fickle and clunky, so be prepared to be frustrated (chapter 8)!

13.1.2 How do I get books onto my Paperwhite?

- **Buy a book from the kindle store:** Buy books from the home screen or the shopping cart button on your Paperwhite. Otherwise login to the amazon store in a browser (chapter 5 and chapter 8).
- **Borrow a Prime reading or Kindle Unlimited book:** If you're a Prime member or a Kindle Unlimited subscriber, you can borrow a book from the Amazon store on your Paperwhite, the Kindle app, the Amazon app, and the Amazon store in a web browser (chapter 5 and chapter 8).
- **Borrow a library book:** Borrow library books using OverDrive's Libby app in participating US libraries (chapter 8).
- **Transfer a book from your computer:** Download a book (free or purchased) and then transfer it to your Paperwhite using sendtokindle (chapter 8).

13.2 Charging

13.2.1 Does ejecting your Paperwhite from your desktop or laptop stop it charging?

No. Your Paperwhite is charging if an orange light displays at its base. The charging icon will also be visible in the top right corner of the screen when you turn it on.

13.2.2 Can you plug the Paperwhite USB cable into an iPhone power adapter?

Yes you can. Any USB charger up to 9 watts will work (chapter 4). Some low power USB outlets (eg. attached to old style keyboards) may not.

I've seen threads on forums saying that some big-brand higher wattage USB-C chargers work, but I have not personally tested any. I've also seen threads that warn against using no-name chargers. So, if you feel comfortable using a higher power USB-C charger, be sure to monitor your Paperwhite closely the first few times you use it.

13.2.3 How do I charge my Paperwhite if my computer doesn't have a USB port?

Use any charger with a USB connector. Most USB chargers will work. The signature edition optionally charges wirelessly.

13.2.4 How do I check the battery percentage on my Paperwhite?

Percentage-charged is shown in the top-right corner of the Home and Library screens, and in the standard toolbar when you're reading a book.

13.3 Subscriptions

13.3.1 Is there a way to read my newspaper subscription on Paperwhite?

You can send news articles from your browser to your Paperwhite (chapter 8). Kindle newspapers and magazines were discontinued in mid-2023.

13.3.2 Do Amazon prime or Kindle Unlimited come bundled with my Paperwhite?

No. These are separate subscriptions. If you sign up to Prime or Kindle Unlimited, make sure you sign up on the same account as your kindle so that your reading benefits flow through.

13.3.3 How many books can you download with kindle unlimited?

The Kindle Unlimited documentation says ten books, but lately I have been able to borrow twenty books at a time. When you reach your borrowing limit, you'll need to return one book to borrow another.

13.4 he Amazon ecosystem

13.4.1 Can you delete your books once you're finished reading them?

Tap and hold the book thumbnail to launch the book menu. From the menu that pops up, Remove download keeps the book in the cloud but removes it from your Paperwhite. Permanently delete removes it from the cloud and your device.

13.4.2 Do you have to be online to access the dictionary?

Initially you'll need to be connected to WiFi to use your dictionary. Once your Paperwhite is setup and settled, the dictionary should be installed locally on your Paperwhite.

13.4.3 Will the kindle book I'm reading on my phone transfer to my Paperwhite?

Anything you're reading in the Kindle app is synchronized to every other kindle device that's connected to your amazon account and connected to Wi-Fi.

When I've borrowed a kindle unlimited book using the kindle app on my iPhone, I've found that I need to open the book on my iPhone before it will download onto my Paperwhite. Also, sometimes you may need to manually sync your Paperwhite (label (20) in Figure 5.3).

13.4.4 Can I gift my kindle to someone with books already loaded?

You can do this, but only if you leave the kindle connected to your account. Kindle's loan-to-anybody functionality was discontinued in August 2022.

13.4.5 Can I lend a Kindle book to someone else?

Anyone with a kindle reader that's connected to your amazon account has access to your kindle library. Different members of my family use different kindle devices, and all are connected to the same account.

Occasionally there's a glitch and I need to go to the Manage Content and Devices area of my account (chapter 10) to Deliver a book to a specific device.

13.5 Hardware

13.5.1 Does my Paperwhite power off automatically if I fall asleep while I'm reading?

Yes. After 10 minutes of inactivity your Paperwhite goes to sleep. It uses very little battery in sleep mode. Some Kindle covers automatically put your Kindle to sleep when you close them and wake it up when you open them.

13.5.2 Why won't the charging cable I ordered from Amazon fit my Paperwhite?

Most likely, you've ordered a charging cable for the wrong model of kindle, or you're using a poor quality or damaged cable. Older Kindles use MicroUSB. 11 th generation and 12 th generation Kindles and Paperwhites use USB-C type cables.

13.5.3 Should I upgrade to a Scribe?

[3 vertical dots] > Settings > Device Options > Storage Management

My decision to buy a Scribe was an easy one. I liked all the features of my Paperwhite, but I wanted to take a step towards making my office paperless. The Remarkable was in my sights. In the end, having a reading tool and a writing tool in the same device was attractive to me. Here's the four things that I considered. . .

- **Interface:** Aside from note taking functionality, the interface on the Scribe and other Kindles is near-identical. That meant an easy learning curve.

- **Hardware:** The Scribe's screen is much larger than my paperwhite's, the battery larger, it has better lighting options, and a faster processor than my old kindle. Also, my 12th generation paperwhite is faster than my 11th generation Scribe.

- **Functionality:** I now have all my books and notebooks with me all the time. I no longer lose notebooks, and my notebooks easily convert to text that I can then fine-tune on my laptop. I also have a diary on my Scribe. And the Scribe has revolutionized my writing workflow. My office is almost paperless because of it.

- **Memory model:** I bought the 64gb model. I usually buy all my devices with as much memory as I can afford because memory can often make a device usable for longer. In saying that, I could have easily got away with buying a model with much less memory.

In the Settings menu, tap Device Info to see how much storage you use on your existing kindle. On my Scribe and my Paperwhite I only use around 1 GB for my books. I could easily have the smallest memory model. My usage pattern - I read mostly kindle unlimited books that I return after reading. I have ~150 books in my library. About equal parts fiction and nonfiction. I do not listen to audiobooks on my Paperwhite. I have lots of notebooks.

If the idea of a paperless office appeals to you, then you should consider buying a Kindle Scribe. I love mine. My Scribe Users Guide (February 2025) shows you how to install and use diaries and notebooks, reading, financial and other tracking documents, crosswords and puzzles. And also my writing and editing workflow.

13.6 General questions

13.6.1 Does Paperwhite use the open dyslexic font?

Aa > Font > Font Family > Open Dyslexic

You access it from the Aa icon in the Display Settings menu (label (31) in Figure 5.4)

13.6.2 Can I use wired headphones?

You can't do this. But you can use Bluetooth headphones (chapter 9).

13.6.3 Can you play games on a Paperwhite?

Sorry. It's books only.

13.6.4 How to get the most from your Paperwhite

- **Use goodreads:** Create a profile in goodreads.com and start creating a list of books that you want to read.
- **Make yourself accountable:** Regularly update the status of the books you've read on the goodreads site.
- **Wish list:** In the browser version of the amazon store, add books to your amazon wish list so you're notified if the price drops.
- **Set a Reading Challenge:** Make public the number of books you plan to read that year.
- **Monitor your reading habits:** Do this from the reading insights area in the home screen of the kindle app on your smart device (look in the hamburger menu). It has information like weeks-in-a-row, days-in-a-row, titles read (by year), and more. Mine looks impressive because my partner shares my account. This information is not displayed on your Paperwhite or in the desktop app.

13.6.5 Does the Paperwhite have a web browser?

Yes. You access it from the 3 vertical dots in the Home screen and the Library screen. It's called an experimental browser, and there's good reason for that. The Paperwhite's processor is not powerful enough to run a browser at the speed that desktops and smart devices can. However, it is convenient for opening website links from a book you're reading.

13.6.6 Is the Scribe waterproof?

No. The Kindle Scribe is not waterproof.

13.6.7 Is Paperwhite waterproof?

Paperwhite has an IPX8 waterproof rating. That means it can handle being fully submerged in one meter of fresh water for up to an hour (Gen 11 and gen 12 specs say 2 meters). So, you can use it in a swimming pool or bath without worrying about it getting damaged. The rating only applies to the device, and not the cable and other accessories.

The rating is intended to deal with accidental submersion. That doesn't mean that you should plan to take your paperwhite scuba diving with you!

The charging port is the weak link in the chain, so make sure it's completely dry before you try to charge it up.

13.6.8 How do I capture a screen?

To capture a screen, touch diagonally opposite corners of your Paperwhite at the same time. You'll see the screen flicker and the screen capture will be saved to your Paperwhite. If the screen doesn't flicker, your simultaneous taps haven't worked.

Screen capture can be temperamental and frustrating. In contrast to the YouTube videos I watched to confirm that the Paperwhite had this functionality, it took me ages to capture just a few images for this book. If you can't get screen capture to work, here's some things to try. . .

- **Fingers too dry:** Try again. It may be that one of your fingers is too dry and the Paperwhite didn't register it.

- **Technique:** You may be tapping and holding instead of doing a quick simultaneous tap. Or, you may be tapping too close towards the center. Try tapping closer to the edge, with your fingers resting on the very corners of the screen – even touching the corner bezel / frame while you're doing it.

- **Restart:** Hold the power button in and restart your Paperwhite (chapter 4)

You transfer screen captures to your desktop computer using the USB cable that came with your Paperwhite. You'll find your screen capture .png files in a folder called Screen captures. From there you can cut and paste them to a folder on your desktop computer as you would with any other file. Be sure to tap the "disconnect" button on your kindle before removing the cable.

13.6.9 What is Goodreads?

Goodreads is an amazon owned social network for book lovers. When its linked to your Amazon account (an option during your Paperwhite setup), you can track your reading progress, rate and review books, and connect with other readers. You can only write reviews from the browser version or the goodreads app on your smart device.

To track your reading progress on goodreads, add books to your currently reading shelf in the About This Book item in the menu that launches from the [3 vertical dots] menu while you're reading a book (label (36) in Figure 5.4) and update your progress as you read. This will allow you to see how far you have progressed in each book, and how much time you have spent reading.

Goodreads also has a social aspect. You can connect with other readers and share your thoughts and recommendations. You can follow other users, join groups with similar interests, and participate in discussions and challenges. All your activity gets shared with your connections.

13.6.10 Can I print a kindle book

No. None of the Kindle apps allow printing, and copyright laws prohibit it. However, you can use the copy and paste functionality in your browser or Kindle app to copy and paste small sections of a book for research purposes.

Chapter 14

Conclusion

Thankyou so much for buying this book and allowing me to show you ways to get the most out of your Paperwhite.

Don't forget the resources page at bit.ly/akpug

If you feel up to doing something incredibly helpful to me, I would be so grateful if you would take the time to leave a review for the Kindle Paperwhite UserGuide 2025 on its Amazon page or on its Goodreads page. Or, if you're feeling incredibly generous, both. I'm an indie author, so I love and I appreciate every reader who takes the time to do that. Every review counts!

Could you spare a moment to review my book?

Thank you so much for reading my book! If you've found it helpful, I'd be incredibly grateful if you could leave a short review. Reviews are a huge help to indie authors. They help other readers discover our books and keep us writing. Just a few words about what you liked would mean the world. Thanks again for your support!

Review on Amazon US	Review on Amazon UK
 https://amzn.to/4hFb80w **Click the Flag or Link** **Scan the QR code**	 https://amzn.to/3EA7u9K **Click the Flag or Link** **Scan the QR code**
Review on Amazon Canada	**Review on Amazon Australia**
 https://amzn.to/3Equs3q **Click the Flag or Link** **Scan the QR code**	 https://amzn.to/4hmKEkv **Click the Flag or Link** **Scan the QR code**

Don't forget the resources from the QR codes in the figures, and at ianallanauthor.com/paperwhite

Made in the USA
Monee, IL
15 September 2026

63198302R00087